Autonomous Ships 2024

Together We Will Shape Our Future

Copenhagen, Denmark
20 - 21 November 2024

ISBN: 979-8-3313-2274-8

Printed from e-media with permission by:

Curran Associates, Inc.
57 Morehouse Lane
Red Hook, NY 12571

Some format issues inherent in the e-media version may also appear in this print version.

The Royal Institution of Naval Architects
8-9 Northumberland Street
London, WC2N 5DA
United Kingdom

Phone: 020 7235 4622
Fax: 020 7259 5912

publications@rina.org.uk

Additional copies of this publication are available from:

Curran Associates, Inc.
57 Morehouse Lane
Red Hook, NY 12571 USA
Phone: 845-758-0400
Fax: 845-758-2633
Email: curran@proceedings.com
Web: www.proceedings.com

160 YEARS OF HISTORY, KNOWLEDGE AND EXPERIENCE

THE ROYAL INSTITUTION OF NAVAL ARCHITECTS

Autonomous Ships 2024

20-21 November 2024
Copenhagen, Denmark

TOGETHER WE WILL SHAPE OUR FUTURE

In Partnership With:

IDA Maritim

THE ROYAL INSTITUTION OF NAVAL ARCHITECTS

Autonomous Ships 2024

20-21 November 2024

Copenhagen, Denmark

THE ROYAL INSTITUTION OF NAVAL ARCHITECTS
8-9 Northumberland Street
London WC2N 5DA

Telephone: 020 7235 4622
Fax: 020 7259 5912

ISSN No: 2752-6569

CONTENTS

Autonomous Ships 2024
20th November – 21st November 2024
Copenhagen, Denmark

MASS – DESIGN FOR INSPECTION

C Balls, Principal Surveyor, Cayman Registry, Cayman Islands

SUMMARY

As the concept of autonomous vessels advances to larger vessels and more widespread international voyages, so the appropriate mechanisms for oversight such as Port State Control (PSC) will need to evolve. It is important that MASS (Maritime Autonomous Surface Ship) design takes the facilitation of effective and efficient inspection into consideration.

This paper will outline some of the discussions that have taken place at Caribbean Memorandum of Understanding on Port State Control (CMoU) Technical Standards Working Group with respect to port state inspection of autonomous vessels, and will include details of development of guidance for PSC Officers, such as a draft aide memoire and will also outline CMoU preparations for PSC activities relating to visiting autonomous vessels in particular with respect to the Cayman Islands.

The paper will also outline some of the issues faced for conducting efficient inspections such as communication, cyber and physical security, vessel access, inspector safety and some remote control center aspects where applicable. Some of the challenges faced by not having international regulations to work with will be discussed. Although this paper concentrates on PSC aspect many of the issues will be in common with flag state and recognized organization inspection, as well as the needs of such bodies as Customs, Immigration and Health Authorities, all of whom may feel the necessity to access MASS to a greater or lesser extent.

1. INTRODUCTION

Shipping is a long established and highly regulated Industry with designs having evolved over many centuries. Automation continues to develop.

The development has initially been with smaller vessels such as Saildrone shown in figure 1.

Figure 1: Saildrones

One area of ship design that needs careful consideration is that of ensuring safe inspection. Conventional manned vessels have the need for provision of safe working conditions for seafarers on board which go some way to ensuring safety of inspectors. Where the need for humans on board to operate the ship is removed then safe access is not required operationally. Most inspection regimes however require human inspectors or surveyors to conduct inspections and surveys. Of course certain aspects of these inspections and surveys can be conducted remotely as well. Cargo surveys may also continue to need humans physically attending

2. INSPECTION REGIMES

With merchant vessels classification society surveys and inspections aimed at establishing the statutory compliance and seaworthiness of a vessel are probably the most common types although some flag states do still undertake these types of surveys and inspections themselves rather than delegating most to class/recognized organizations. Cargo and insurance surveys are also undertaken as are vetting inspections.

This paper will concentrate on challenges faced with Port State Control Inspections which are essentially the safety net introduced for Port States to ensure that vessels entering their ports are of a suitable safety standard and identify issues that may have been overlooked by the surveys audits and inspections conducted by or on behalf of a vessels flag state.

3. PORT STATE CONTROL

Port State Control (PSC) is the inspection of foreign ships in national ports to verify that the condition of the ship and its equipment comply with the requirements of international safety and environmental regulations and that the ship is manned and operated in compliance with these rules.

Many of the International Maritime Organization's (IMO) most important technical conventions contain provisions for ships to be inspected when they visit foreign ports to ensure that they meet IMO requirements. PSC is also covered in the International Labor Organization's (ILO) Maritime Labor Convention. These inspections were originally intended to be a back up to flag State implementation, experience has shown that they can be extremely effective.

The IMO adopted resolution A.682(17) on regional co-operation in the control of ships and discharges promoting the conclusion of regional agreements so that PSC inspections are coordinated in order to focus on substandard ships and to avoid multiple inspections. IMO Resolution A. 1155(32) Adopted on 15 December 2021 covers the procedures for port state control.

Nine regional agreements on PSC - Memoranda of Understanding or MoUs - have been signed:

- Europe and the North Atlantic (Paris MoU);
- Asia and the Pacific (Tokyo MoU);
- Latin America (Acuerdo de Viña del Mar);
- Caribbean (Caribbean MoU);
- West and Central Africa (Abuja MoU);
- Black Sea region (Black Sea MoU);
- Mediterranean (Mediterranean MoU);
- Indian Ocean (Indian Ocean MoU);
- U.A.E., Bahrain, Saudi Arabia, Oman, Qatar, Kuwait (Riyadh MoU).
- The United States Coast Guard maintain the tenth PSC regime.

The regimes all operate in similar manners, however there are some variations and also the enforcement of regulations is dependent upon which conventions are adopted by the particular administration, which in turn may affect the types and sizes of ships inspected. Similarly, the reporting forms used whilst similar have differing formats. All regimes have manuals to assist their inspectors with the utilization of checklists and aide memoires in many cases.

The inspection targeting system used varies but, in all cases, aims to prioritize those vessels most likely to be substandard.

Whilst the inspection processes and definitions do vary slightly amongst the various regimes in general there are four main types of inspection;

Initial Inspection which involves a visit on board a ship by a port state control officer (PSCO), in order to check compliance with the relevant conventions and regulations and verifies, where appropriate, whether outstanding deficiencies, found during the previous inspections carried out, have been rectified and satisfies himself of the overall condition and hygiene of the ship. Certificates and documents will be checked. In circumstances where clear grounds are found this inspection may be escalated to a more detailed inspection.
More Detailed Inspection. which consists of a more detailed inspection in the area(s) where clear grounds were established, as well as potentially in other areas at random. It could include further checking of compliance with on board operational requirements by carrying out operational controls and drills.

Expanded Inspections are sometimes conducted on high-risk ships and includes a check of the overall condition, including human element where relevant and will include the following risk areas, subject to their practical feasibility or any constraints relating to the safety of persons, the ship or the port: As with all types of inspection the PSCO will use professional judgement to determine the appropriate depth of examination or testing of each specific item.

Overriding Priority (additional) Inspections mean that an inspection is conducted earlier than would normally occur under the targeting scheme in use and can be triggered by several issues, including such things as Seafarer and Pilot complaints.

Concentrated Inspection Campaigns are also carried out from time to time where a certain area of inspection is addressed in more detail and a specific checklist completed to allow effective analysis of the results. At some stage in the future such a campaign could be carried out with respect to autonomous and remotely controlled vessels, although this is not currently being considered.

Focused Inspection Campaigns (FICs) aimed at improving the effectiveness of inspection campaigns have also now been carried out on a trial basis and are likely to continue to on announced and unannounced basis.

4. THE CARIBBEAN MoU on Port State Control

Port State Control Aide Memoire Autonomous Vessels < 24m

On Quay with agent/owners representative				
Certificates/Documents available (see Annex)				
Certificates/Documents valid in form and content (electronic certification validated as per FAL.5/Circ.39/Rev2)				
Documents for cargo carried according to CoF/IMDG/IMSBC				
Remote Control Centre verification document				
Remote Control Centre operator certification				
Remote Control Centre operator works schedule and hours of rest record				
Remote Control Centre operator medical certificate				
Management of charts and publications/ECDIS evidence				
Passage Plan				
If applicable Emissions related records				
Previous Port State Control Records				
On Board vessel with agent/owners representative				
Deck				
Navigational equipment (proper operation V/16.2)				
Cyber security arrangements				
Test of steering gear				
Lifesaving appliances – e.g. Lifebuoy - recovery ladder				
Navigation lights and signals including light and sound signals				
Anchoring arrangements				
Cargo Handling				
Cargo securing devices				
Cargo gear and additional equipment				
External Hull				
Load-line and draft marks				
Hull condition – indents etc.				
Guardrails and access arrangements as applicable				
Mooring arrangements				
Emergency towing arrangements				
Deck condition				
Hatch covers and seals				
Ventilators				
Communications equipment, aerials etc.				
Engine Room/ machinery Spaces/ Steering Gear				
Overall cleanliness				
Pipe work				
Bilges Clean				
Bilge pumps and valves				
Fixed fire extinguishing system				

Generators, Switchboards, and presence of earth faults				
Lighting				
Cables, Cable Glands, Terminations, deck and bulkhead penetrations				
If fitted; Oily-water separator, equipment, and Oil Record Book				

Note; Where communication with remote operating centre is needed a reasonable time should be allowed for the remote operator to arrange cover for other vessels that they may be responsible for so that they are not distracted from operating other vessels safely by the ongoing PSC.

Figure 2: Draft Aide memoire for initial inspection of small autonomous vessels

The CMoU Technical Standards Working Group met in February 2024 and discussed the approach to autonomous vessels particularly with respect to remote operated vessels under 24m L as this is an area where activity is already happening in the region.

Discussions took place with respect to risks posed by these vessels. When there are no seafarers on board then there is no direct risk to their safety however of course the interaction with manned vessels does potentially present risk to human life.

The environmental risks are considered similar between both conventional and autonomous vessels.

Various potential approaches were discussed such as requesting of DIHMAR (Directors and Heads of Maritime Administrations – Caribbean) that at the next Caribbean Small Commercial Vessel Code (CSCV code)[1] update it has an annex added regarding autonomous vessels similar to UK's workboat code However possibly a Caribbean remote controlled vessel code for vessels under 24m L could be developed as a better solution provided its review schedule aligns with that of the CSCV code so there is no regulatory drift between what would effectively be separate codes for conventional vessels and autonomous/remote controlled ones. At this stage it would not be envisaged to tackle fully autonomous vessels but just remote controlled ones. By the completion of the next CSCV code review the new IMO autonomous ship code should be in place so terminology etc. can be aligned.

The Caribbean Cargo Ship Safety Code (CCSS Code)[2] is currently nearing completion of its 3 yearly review cycle so tackling autonomous vessels in that will be left until at least the next review which will still be before any mandatory IMO code is introduced

In the meantime, an aide memoire for inspectors of autonomous ships is under development (see figure 2) and trial which covers the main aspects of autonomous ships applicable to vessels under 24m and includes basic aspects of remote control centers as well as the appropriate parts of the SCV code. This aide memoir was first used during the Port State Control Inspection of an autonomous vessel on 24 August 2024 in George Town Grand Cayman.

As such areas as the use of pilots do not usually apply to these smaller vessels they have been omitted. The possibility of carriage of passengers will need consideration as there is of course potential for such vessels as harbour foot ferries as has been seen in European developments.

5. THE REQUIREMENTS OF INSPECTORS

Port State Control Officers (PSCO) in conducting their duties expect to have safe access to a vessel and also be able to interact with the seafrers on board. As part of the overall safety considerations for these inspections they expect to be accompanied on board at all times.

If a vessel is unmanned then of course there are no seafarers available to accompany the PSCO. That said it is likely that the vessel will have to have some kind of human representative in port similar to the current ships agents' arrangements, however the requirements for ships agents are not really regulated and of course the current arrangement is that they provide communication between the ships staff , the port and relevant authorities and the ships mangers/owners. Thus, the role of the ship's agent will be slightly different for autonomous vessels and some form of training will need to be developed to align their skills with those necessary to effectively deal with the needs of an autonomous vessel in port.

Security wise as there are no persons on board then many spaces can be secured and do not need access except for periodic maintenance. On a conventional ship the PSCO can just ask for a sample of spaces to be unlocked so they can see inside to check there is nothing untoward (Such officials as customs and immigration officers have similar needs in performing their functions). Thus, a means of unlocking spaces needs to be provided whether remotely or by a human representative, it is likely that the human option for this will be preferred by owners and operators certainly in the early days. PSCO's could be given remote familiarization training to ensure safe access to a particular autonomous/remote controlled vessel and their presence on board could be monitored by the remote operation center through camera systems. Such training will of course be needed for any persons gaining access to this type of vessel.

PSC reports use codes for statistical analysis and these codes will need updating to include aspects of remote-controlled or autonomous operation such as any weakness found in remote control centre communication etc.

With conventional ships a copy of the report is left with the master but in the case of remote controlled or autonomous vessel such report would probably be sent electronically to one of the control center's or the vessel operators office

Whilst many types of survey and inspection are planned and scheduled with arrangements agreed with the owner, PSC inspections are usually partially random inspections without preplanned arrangements made with the owner or indeed the vessel in order to maintain an element of surprise and thus assess the vessel in a reasonably true operating environment.

As has been previously mentioned such authorities as customs and immigration will also be needing access to autonomous vessels to check for such things as contraband and illegal immigrants, it will be beneficial for such visits to be coordinated with port state control activities in order to assist with maintaining good security levels as well as enabling an element of surprise to still exist with these visits to assist with regulatory enforcement.

Figure 3: Inspection hatch open on Saildrone 3001

6. THE REQUIREMENTS OF OWNERS

Vessel owners will want to maintain the security of their asset and restrict both physical and Cyber access given to other parties.

As with all types of security the level of protection will vary according to the perceived threat.

From a physical security point of view the camera system used for maintaining lookout can be utilized for spotting security risks at a distance, however the close up security of the vessel will probably need monitoring by a separate system although potentially using similar automated target recognition technology.

From a Cyber Security aspect it would be reasonable to aim for a similar level of protection as is applied to the financial services industry to allow international transfer of significant monetary funds.

From an anti-piracy point of view appropriate levels of security need to be maintained but there must also be proper arrangements for emergency towing.

7. CONCLUSION

The safe conduct of inspections on autonomous ships presents some different challenges to those with conventional ships, however these are not insurmountable.

Provided autonomous systems are designed to allow interrogation by the appropriate authorities then the effective oversight of the operation of autonomous ships can be achieved, this will be much easier if common international standards are used and appropriate mechanisms for the assurance of autonomous systems are developed.

Inspection techniques will also need to be updated to make effective use of new technologies and not rely entirely on traditional methods

8. AKNOWLEDGEMENTS

The aforementioned IMO conventions and related texts and CMoU meeting reports/procedures.

Saildrone inc. for Figure 1

9. REFERENCES

1. Caribbean Small Commercial Vessel Code August 2021 prepared by Caribbean countries with assistance of the International Maritime Organization.

2. Caribbean Cargo Ship Safety Code October 2020 prepared by the Caribbean Memorandum of Understanding on Port Sate Control.

10. AUTHORS BIOGRAPHY

Chris Balls is a Principal Surveyor with the Cayman Islands Shipping Registry. He is a Master Mariner who spent 25 years at sea before coming ashore as a Marine Surveyor with the UK Maritime and Coastguard Agency for ten years and then spent four years as Chief Marine Surveyor for the Antigua and Barbuda Department of Marine Services and Merchant Shipping before moving on to Cayman Maritime as Principal Surveyor in the George Town Headquarters, with responsibilities including ensuring the regulatory compliance of the country's fleet and visiting vessels. He is a Fellow of the Royal Institute of Naval Architects and Associate Fellow of the Nautical Institute.

THE PROFESSIONAL PROFILE OF THE REMOTE OPERATOR

C Plawenn-Salvini, University of Applied Sciences Emden-Leer, Germany
O T Gudmestad, Western Norway University of Applied Science, Haugesund Branch and UiT, The Arctic University of Norway, Tromsø

SUMMARY

The need for experienced nautical officers is imminent, the possibilities to work ashore which requires their expertise are plentiful. Now, a new maritime profession is emerging, the remote operator. Again, we require mariners to take that role, resulting in a further increase in the personnel shortage. It is time we consider an alternative to sate the need for qualified persons in the remote operation centers. Before developing a new maritime education for shore-based personnel, we need to understand the professional profile of the remote operator. This paper discusses the different modes of remote operations, what tasks are expected of the remote operator, what other duties await them, risks, challenges, and what they can do during shipboard emergencies. Finally, we summarize the suggested competence requirements for remote operators.

1. INTRODUCTION

It is human nature to jump ahead, at least five steps at a time. We observed the same with the development of MASS (Maritime Autonomous Surface Ships). 'If nobody is on board what happens if there is a fire?' such questions were posed in the conceptual phase of YARA BIRKELAND. The IMO (International Maritime Organization) offered a tool to create a common understanding of concepts and ideas; the four degrees of MASS [1]. It is commonly accepted that Degree One represents the standard of autonomy on conventional ships nowadays [2]. DP (Dynamic Positioning) Systems, ECDIS (Electronic Chart Display and Information System), ARPA (Automatic Radar Plotting Aid), Track Pilot, and even the Autopilot all represent automated systems that support mariners in their watchkeeping routines.

We understand today that the four degrees are too rigid a system, still, it was important to have them initially to draw a pathway in the development of maritime autonomy. Shipboard operations may run fully or party autonomous in one department and conventionally in another. The boundaries have become very soft and may be stretched or indented to fit the needs of a certain application.

Seeing the development of autonomous shipping and the current trend with the commissioning of ROCs (Remote Operation Centers) it seems they will soon start to sprout like mushrooms. To man these with competent personnel, REs (Remote Engineers) and ROs (Remote Operators), fully trained maritime professionals are employed, preferably with years of seagoing experience. This approach ensures a thorough understanding of vessels, ship handling, stability, etc. is readily available. Additional training for remote operations and in-center-time sees them qualified to work in the ROC. However, if we look at the local maritime labor market, there are many more vacancies than applicants. This is reflected in the graphs in Figure 1. In only three years the number of masters and officers holding a valid European CoC (Certificate of Competency) has dropped by approximately 10 %, while the number of vessels worldwide has risen by 6%.

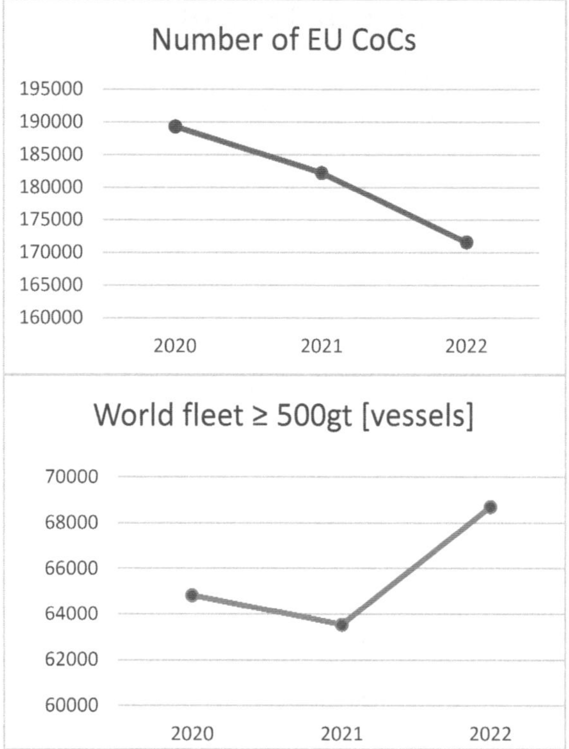

Figure 1: Comparison of European CoC holders [3] versus the number of vessels above 500 GT (Gross Tonnes) [4]

The graphs highlight that the maritime industry in Europe is struggling for mariners. This is not just an interpretation of numbers, but also an opinion by many relevant players. 'Sweden has gone as far as opening an additional course to train nautical officers, without success, as there were zero attendants,' stated one of the interviewees for this study. Further, only a part of the persons completing their maritime education are working onboard to get valuable seagoing experience, which is needed in all areas of the shipping industry. The struggle to find qualified personnel for conventionally crewed vessels may even be one of the major drivers for autonomous shipping. The same effect is already seen with inland waterway traffic which is presently changed by SeaFar through their agenda to operate remote-controlled vessels [5].

The approach to utilizing certified shipboard crew entails a big problem though. As mentioned above, we already have a shortage of qualified personnel and now we want to take another chunk out of that group and put them in the ROCs. However, autonomous shipping will not replace traditional shipping and the need for mariners. There is even the argument that it will enable companies to go small, meaning that there is not one big vessel with maximum capacity, but several smaller vessels that offer more flexibility in their deployment. They may be manned with a reduced crew and can create more flexible schedules, but still cover the peak times sufficiently. 'We will not replace a good colleague with a robot' was stated by an interviewee. The want for seafarers means that autonomous shipping can be seen as an augmentation to traditional shipping, as a supplement. It will not be a replacement, as an unmanned commercial vessel crossing the oceans is still unviable [6].

It should be a priority for the industry to counteract the increasing discrepancy between available nautical personnel and the needs of the labor market. An example of failure to achieve this is the certification of DP Operators which led to a considerable bottleneck in the crewing of offshore vessels. A dedicated education for remote operators appears to be the logical solution. That does not mean mariners should not be able to pursue this career path. Training them will be valuable and likely they can gain seniority more quickly than a dedicatedly trained RO. But a direct route to the profession should still be available, without mastering another one first.

To avoid the common pitfall of jumping five steps ahead and already designing an educational program for laymen, the first step on this path is to outline the profile of the RO. We need to understand the work environment, primary and additional duties, risks, and challenges of the occupation to get a clear picture of this novel profession, how they work, where they work, and what is expected of them.

To achieve this, a series of qualitative interviews have been conducted with leaders in the industry, education, and certification. The interviews were conducted in a very open manner with only some guided questions regarding the approach to autonomous shipping, and the role and expectations of their personnel, both in everyday and in emergency situations. To analyze the interviews, the method of systematic text condensation was applied in which the topics were reviewed, further fragmented, and sorted. The statements were then condensed and compared for commonalities and differences [7]. This shall ensure that the professional profile of the remote operator is not just a theoretical construct, but reflects the reality as it emerges and already exists.

2. THE WORKPLACE OF THE REMOTE OPERATOR

Shipping companies have invested a lot into automation and information gathering, making more advanced systems commonplace. DP has for example experienced a boom within the last decades, but most traditional companies are hesitant to take the step towards autonomy. Still, the advances in sensor techniques and centralized data processing of data, have provided the possibility of vessel autonomy. However, it is primarily young companies or purpose-founded joint ventures, that endeavor upon this new path. They mostly place themselves within one market segment, i.e. ferries, ro-ro vessels, offshore operations, etc. If that does not hold, they at least stick to a similar technological solution for their ships, and the ROC. The vessels' work scope remains the same though, it just so happens that the manning has shifted to some degree from ship to shore. Diversifying into different market segments or various technology providers early seems to be too great of a risk since both technologically and legislatively we are still in a touch-and-go phase.

Outsourcing the crewing needs of the ROC is not an option for these new players in the industry. Rather they see themselves as a service provider to companies looking for innovation, offering remote crewing and ship management capabilities. With considerable success, for example, ASKO, YARA, and REACH SUBSEA, which have not previously been engaged in ship management, are driving forces in developing autonomous shipping by contracting MASSTERLY [8].

In general, we can distinguish three modes of remote operations:
- Supervision of autonomous systems
- Remote control of vessels
- Remote navigation support

2.1 DIFFERENCES IN MODES OF REMOTE OPERATIONS

Supervision of autonomous systems is the highest degree of autonomy one strives to achieve. A vessel departing, transiting, and docking by itself is an intriguing idea. These vessels are supposed to run on routes defined by the RO, be capable of acting according to COLREG (Convention on the International Regulations for Preventing Collisions at Sea) [9], and maneuver in port by themselves. The role of the RO in this context is multi-tasking orientated, as they will have to keep an eye on multiple vessels simultaneously. One company is talking about one operator supervising five vessels, another would like to go up to as many as eight, however, they agree that only time and the level and reliability of the autonomy will give a clear picture of the true capabilities. It will be the duty of the RO to recognize situations that require human interference and to troubleshoot in case of failure or breakdown. However, there is no direct control over the vessels. The RO may define waypoints where the vessel should be heading, like a track pilot, or put the vessel into its fallback state, in which it is attempting to maintain position and await further instructions. Correctly identifying the situations when intervention is required will demand a lot of focus and immersion into different situations simultaneously. Therefore, time in the chair while observing a number n of vessels, is heavily limited. Supervision of autonomous systems is always done in teams, for now, it is assumed that a team consists of two ROs and one RE. Repeatedly rotating the two ROs in a shift ensures the immersion-heavy nautical situational awareness is limited in time. If a situation is deemed to exist, which requires more focus from the RO, there will always be the second person who can supervise the $n-1$ remaining vessels. The application of these vessels will mostly be in short trade. Fixed routes and limited distances are an ideal environment. The situations represent relatively simple traffic situations and the restricted area will quickly be familiar to the RO.

Figure 2: A workspace for supervision of autonomous systems, MASSTERLY's ROC in Horten, Norway

Remote control of vessels is the direct control over a ship from an ROC, either via a joystick or a conventional setup of controls. This method has been used for some time in the operation of USVs (Unmanned Surface Vehicles) as they are small enough not to fall under SOLAS (International Convention for the Safety of Life at Sea) regulations. Operation of such vessels then falls under national legislation, which is easier to change and adapt, compared to the IMO rules. The same applies to inland waterways, where cargo vessels are already being remote controlled. Crew costs and personnel shortages are higher compared to international shipping [10]. In recent years this method has seen a respectable expansion in Europe. One operator exclusively handles one vessel at a time. This approach may also be used at sea temporarily, when in low-traffic areas, to reduce work hours for the onboard crew or when there are not sufficient navigators onboard to operate on a 24/7 basis. The RO and the OOW (Officer Of the Watch) must conduct a detailed handover before the controls may be passed between the ROC and the shipboard crew. Remote control of vessels might also be used when the bridge crew is supposed to be engaged with other duties. For example, on conventional ships, one of the bridge officers cannot be involved in most emergency drills because someone must control the ship. Further, compliance with resting hours may be possible, even in high-frequency port rotations, which at present is often only the case on the records.

Remote navigation support refers to operations with a manned bridge, where the RO takes an advisory or supplementary role to assist the OOW remotely from ashore. This person can log onto a bridge, access relevant navigation instruments, and use a range of cameras. The lookout duties during nighttime or when passing through dense traffic areas can thus be done from the ROC and free up capacities onboard. A company following this remote navigation support approach stated during the interview that these duties are still fulfilled mostly by senior and some junior nautical officers. The reason is that the technology may be updated, and processes and features can be optimized. Another benefit is that they are using the captains of their fleet in the ROC; they will spend some time in the operator's chair and then return to their deployment at sea. This approach is on the one hand building trust in the ROC and the remote operator, on the other hand, it allows the mariner to see both sides of the operation and this knowledge can in turn be used to update the system. In the offshore industry, it is not uncommon to have client representatives onboard, this necessity is made obsolete, as the representative may observe the vessel's position and actions from the ROC. This will reduce the distraction for the bridge crew, and minimize the representative's time who only must drop by the ROC during relevant times and not spend the whole span of the deployment on board.

2.1 COMMONALITIES IN MODES OF REMOTE OPERATIONS

Despite the differences in modes of operation, we should not fall back into the habit of creating rigid systems where something can be either one or the other. Instead the transitions need to be seen as flowing. Presently, companies are operating either one mode or another, but there is potential to vary a vessel's mode of operation throughout one voyage. In a hypothetical scenario a ship may be departing under manual control, and an OOW takes the navigational watch alone. Due to darkness, restricted visibility, or density of traffic, he asks for remote navigation support. When the lookout is no longer needed, he will relieve the remote support navigator and later, during the deep-sea passage, will switch the system to autonomous with appropriate supervision. Another example is Reach Remote, a vessel with DP capabilities, but without DP notation, navigating to the offshore field autonomously with supervision and then being remote-controlled whilst working with ROVs or conducting survey works in DP-like operations [11]. These two scenarios highlight that the different modes of remote operations are closely related – be it in conjunction with manned or unmanned vessels.

One big challenge for all modes of operation is the flow of information. Modern vessels are equipped with thousands of sensors, that produce a lot of alarms. The flood of alarms is already very much an issue on conventionally crewed vessels [12], if the same continues in the ROCs, it will make it difficult to maintain situational awareness for a remote-controlled ship, not to mention several supervised autonomous vessels. The same also applies in the other direction, a risk of the remote navigation support is to announce too many sightings that are not imminently relevant, leading to a distraction of the OOW and thus deteriorating the situational awareness. Reducing the information flow to the essentials is a prerequisite to achieving a reliable level of situational awareness necessary for competent and safe navigation.

The vessels handled by the RO will vary concerning the crew on board. As indicated earlier, we are moving away from the rigid four degrees of MASS and are looking at more flexible and individual solutions. However, the classical need for familiarization with the vessels must remain a high priority. This can be done by spending some time on board, in case there are still crew accommodations available, or by just visiting the vessel. But even VR (Virtual Reality) is already in use, which offers a complete walk-through of the vessel, including the LSA equipment, as stated by one company. This is particularly useful when the vessel is operating worldwide and there is no chance to physically visit the vessel, without excessive effort.

Something all players agree upon is that when the crew is onboard, it will always have authority over any remote command, the exception is a security protocol that may be engaged in case of hijacking events.

Another requirement for competent remote operations is radio communication. Hailing other vessels on VHF (Very High Frequency) radio, broadcasting Sécurité messages during surveying or towing, and receiving navigational warnings or distress calls are fundamental for safe ship operations. For short trades, a restricted operator's certificate would satisfy the requirements, but for worldwide trading, the ROs will have to comply with the general operator's certificate.

3. ADDITIONAL DUTIES

Looking at the three modes of remote operations, it becomes apparent that the RO will have some time available apart from their main task, i.e. partaking in the vessel's safe transit. In the case of remote control and remote navigation support, those times are when the RO is not actively engaged with a vessel, and in the supervision of autonomous systems, when the vessels are behaving as they should, the second RO has time to engage in other tasks.

Operating a vessel entails much more than just bringing her from A to B. Routes must be planned thoroughly from pier to pier and may need adjusting according to meteorological conditions. Passage planning is something that all players agree, can be done by the RO, not necessarily only for the vessel under one's immediate care, but also for any vessel within the fleet.

Further, the duties of the RO will entail the scope of a regular nautical officer, planning and conducting maintenance, scheduling services, organizing the renewal of certificates, and ensuring that the vessel is upheld and in good working condition. The advantage of the RO taking over these duties is that they are in a hybrid position of ship and shore, effectively making the interaction between the officer and the vessel manager or superintendent obsolete and challenging the classical hierarchy of a shipping company. However, it will likely not be every RO who can place orders, instead, there will still be a person in charge, who may also be fulfilling the role of a RO. Some companies strictly separate ROC personnel from the rest of the office staff, due to the vessel's complex deployment planning and the requirement to not distract the ROs in their duties. However, a certain level of understanding of planned and condition-based maintenance, organizing services, and renewing certificates may be accepted within their work scope.

Several companies have also indicated that it would be useful to have the supervisors of autonomous systems acting as a VTS (Vessel Traffic Service), surveying and controlling the ship traffic within their area of operation, in addition to the RO's regular duties. Especially for the supervision of ferries, this makes a lot of sense, as they are trafficking repeatedly over a certain stretch of water, which the shoreside crew will be well familiar with. Anticipating the departure time of the ferry and the resulting ship encounters, it seems self-evident that it would not only be possible for the RO to cover such duties but also be beneficial to their situational awareness. Developing close-quarter situations can be spotted early and vessels will more likely adhere to instructions received by a VTS, than a ferry's request to give way. An undertaking aiming to include autonomous ferries into the public transportation network further stated that the RO can function as a traffic manager so that bus and ferry schedules are coordinated to provide smooth transitions for the customers.

None of the interviewees have stated that stability is amongst the duties of an RO, however, they clarified the reason being that either the vessels do not even have a ballast system, or if they do, the stability is always more or less the same. There are not sufficient activities involved for the shipboard crew to justify a remote stability computer or remote ballasting system. However, this does not mean that verifying stability would not be part of the RO's duties in the future. A basic understanding of buoyancy, stability, dangers regarding the lack thereof, and resonance effects of ships must be considered essential for overseeing a seagoing vessel.

4. EMERGENCY PREPAREDNESS

Talking to the partners in the industry has made it clear, that the initial argument of developing autonomous shipping operations to promote safety has moved to the background. The financial aspects dominate the strive for innovation. Safety is a prerequisite to achieve a profitable business case. However, the mantra holds that operating autonomous ships must be at least as safe as a conventionally crewed vessel.

The approach of contingency planning varies significantly. However, there are some commonalities, like the argument that the new vessels are already designed to be safer, thus the threat of fire is regarded to be accounted for by MASS with their fixed firefighting systems. For example, on small work boats, the crew would not conduct firefighting in any case, so the new designs with state-of-the-art water mist and inert gas systems increase onboard safety.

In the case of partly crewed or passenger vessels, it is possible today to launch life rafts and evacuation systems from the bridge. The crew must be familiar with these systems and catering or service staff would still be available to assist passengers. There is then little argument that LSA (Life Saving Appliances) should not be launched from the ROC instead of the bridge.

Assisting in an external emergency, like aiding a casualty in the water is considered fulfilled. According to UNCLOS (United Nations Convention on the Law of the Sea) Article 98, there is an obligation to render assistance [13]. The method of how to render the assistance is not further defined. The interviewees agreed that during SAR (Search and Rescue) activities MASS will likely not be retrieving persons from the water but instead be utilized to create a lee, assist in searching using advanced sensor arrays, feed information and potentially even video to the MRCC, and take the role of the OSC (On-Scene Coordinator). Special training should be provided to prepare a RO for such eventualities. Despite these opinions regarding rendering assistance, the current draft of the MASS Code states *"Mass should have [...] specific plans, procedures and training and drills for the rescue of persons in distress, [...]"* [14]. If such a requirement is implemented in the mandatory MASS Code, companies such as Zelim, offer remote-controlled fast rescue boats [15], which enable also partly or completely unmanned vessels to comply.

Another concern by players in the industry is still cyber security. Arguably, it should not only be a concern for the companies in question but also for all seagoing and port personnel, since a vessel being controlled with malicious intent may cause much harm. ROs must be trained to recognize unwanted interference and to provide appropriate responses to such events.

For a short trade of less than one nautical mile, it is being argued that the best response to internal emergencies is always to reach the closest port. However, this should not relieve them of contingency training. The players agree that any training may be conducted using simulators, or in joint exercises with the shipboard crew. Emergency drills are essential to maintain awareness of emergencies and correct behavior, this applies to traditionally manned and autonomous vessels.

5. CHALLENGES

Of course, there are still a lot of challenges ahead regarding autonomous shipping. We have seen that remote control of vessels works well inland where rough seas, complex multi-vessel traffic situations, and connectivity are less of an issue. However, the traffic is busy in coastal areas, which makes it more difficult to achieve full situational awareness, while in the open ocean, connectivity might be more of an issue, as Japanese test runs have reported occasional problems during field tests [16], also there is an increase in data traffic expected, once the non-mandatory MASS Code will be completed [17] with yet to be seen consequences on the available bandwidth.

Mainly legislative issues are restricting the full potential of autonomous shipping though. Currently, national solutions are being sought and found, and as soon as there is sufficient proof of concept these national solutions can grow into bilateral and multilateral agreements before finally being implemented within the IMO.

In the meantime, there are still challenges to be overcome regarding educating and training the remote operators. A considerable issue in training is to achieve the situational awareness that is inherently adopted by mariners when looking outside the bridge windows. At the same time, the introduction of technology will make it easier to identify objects, judge distances, and even get offered suggestions for positive action. Generically trained personnel, who only know this way of involvement and decision-making, might work more efficiently when using these systems, compared to conventionally educated mariners with an extra RO's training on top.

Nevertheless, it is still likely a good idea to have at least some sea-time experience in a dedicated RO educational program, as putting trust in someone half a world away, without having ever met them, must be well founded. Globalization, and the COVID-19 pandemic, have made cooperation between teams spanning different continents commonplace. When the connectivity is ensured, this is also possible for shipboard operations. However, cooperation onboard ships is very different from many other sectors. The shipboard crew is putting their lives into the hands of the person or the system in charge of the navigational watch. A level of trust is required, which must be built within the bridge team and in the entire profession. Initially, it may be easier to trust someone who you know has received the same training and education you have as a crewmember onboard. Achieving this confidence is a considerable hurdle for future dedicatedly trained ROs, without being a mariner first. Skepticism may be alleviated by spending at least a limited amount of time on board during the education.

6. AVAILABILITY

There is not necessarily a lack of persons interested in working in the maritime sector, but the life of a seafarer is not always easy. Some examples are extended periods far away from friends and family, long working hours, isolation at sea, and confined living quarters. Many drop out early from the education schemes for those reasons, others finish but never sail as an officer. The occupation of the RO opens the possibility of pursuing a maritime career without these inconveniences. Offering two paths, instead of one may be highly beneficial to counteract the personnel shortage. Also, the retention rates might be much longer. If seagoing personnel suffer from a serious illness or an accident, they likely lose their medical fitness certificate. Further, in Europe, females only represent 2.5% of the total CoC holders [18]. They immediately lose their medical fitness certificate when pregnant and after giving birth, the reintegration into the shipboard profession, potentially part-time, is difficult to accomplish. The work in a ROC can accommodate the above needs more easily, offering less restrictions and better chances for reintegration into the labor market [19] leading to a more diverse workforce.

The interviewees also agreed that the ROC crew would not earn less than their shipboard equivalent, likely even slightly more. Whether that is only a snapshot view of the early days of ROC existence or it will be a long-term trend, remains to be seen.

Summarizing, it can be said that ROs will have a more socially oriented work environment, better suited to accommodate diversity. Offering a dedicated education will enable people to work with ships in an interesting environment without being trained as mariners first.

7. SUMMARY OF COMPETENCE REQUIREMENTS FOR REMOTE OPERATORS

Large sums are being invested into ROCs, and the proof of concept will be when not only remotely controlled and supported navigation is viable, but also when safe operations and supervision of autonomous systems are achieved. On the education side, we must recognize and understand this trend, to provide appropriate training schemes for the new profession of the RO.

Summarizing the competencies that are required according to industry leaders provides us with the following list:

- **Basic understanding of vessels** can be accomplished through familiarization in conjunction with onboard experience. The practical training will help to build trust with the crew still onboard. Also, it will avoid detachment from the vessel when making maneuvers. The first author has frequently observed nautical students, without previous bridge experience, making unnecessarily hard maneuvers in simulator training. In real life, this is not only uncomfortable for persons on board, but may even be dangerous to the ship and the cargo.
- **Ship handling** at least to a limited extent, as the amount of direct control will be heavily limited. Whether the current approach of only dedicating waypoints will prevail remains to be seen. Alternative solutions could be joystick control or a limited range of emergency maneuver options.
- **Sensor fusion** is performed by autonomous systems, meaning that the information from different sensors is being overlayed and combined. A thorough understanding of the different systems and their limitations is the key to a correct interpretation of the displayed data.
- **Situational awareness** is the cornerstone of making autonomous shipping work. This will require not only an understanding of the above-mentioned sensor fusion but also extensive simulator-based training.
- **Coordination of traffic situations** is essential for maintaining situational awareness of multiple vessels simultaneously. The ROs can in turn be used to act as a VTS or act as an OSC during SAR activities.
- **Familiarization with the trading area** needs to be addressed, including knowledge of local navigational hazards, expected traffic, and surrounding conditions.
- **COLREG and national rules of the road** must be adhered to. Knowing when and how to react during vessel encounters is key for every manned and unmanned ship.
- **Route planning** may be conducted by ROs when not directly engaged in navigation or supervision.
- **Inspection, certification, and maintenance planning** are other duties for idle periods.
- **Resilience and emergency response training** must be part of both the education and regular contingency training. A takeover during emergencies may be needed and the RO must be competent to ensure the safety of the vessel under critical situations.
- **Cyber security threat recognition and response** need to be schooled, trained, and updated regularly.
- **Radio operator** is a qualification which will enable the RO to correctly convey information and intentions. Knowing and applying correct radio procedures is the foundation for many of the above-mentioned points.
- **Stability** is essential for any vessel of any size. A base understanding of risks and dangerous resonances is a must for any person in charge of conducting a watch, whether onboard or ashore.

8. CONCLUSIONS

It is important to understand that the stated required competencies are mainly reflections of the industry players. Of course, these cannot be the only source, but their insights are relevant to creating an educational scheme for the RO.

Autonomous shipping is an exciting new field in the modern maritime industry which opens the door for a young and more diverse workforce. The different modes of remote operations may be defined as supervision of autonomous systems, remote controlling of vessels, and remotely supporting a navigational watch. In the future ROCs may offer the entire range of remote operations within their fleet or to their clients, so certified personnel should be trained in each operation accordingly. While these applications differ, there are still significant commonalities, making a common education viable. The required trust by the remaining crew and passengers onboard must be built through a comprehensive education of the RO, which includes at least a limited amount of onboard training.

The further development of autonomous vessels will enable the industry to shift from maximum capacity vessels to diversifying and deploying smaller units. To manage these vessels, we need to expand the workforce. A designated educational program for remote operators will lay the foundation for traditional and autonomous shipping to unfold their full potential.

The challenge of defining key competencies and how to involve them in educational programs are being addressed by various research groups [20, 21]. This paper in contrast focuses less on the theoretically needed skills, but on the requirements of the industry. To avoid a tunnel vision of the industry's pioneers, further interviews will be conducted with companies not yet engaged in autonomous shipping, to compare their expectations with the findings of this paper.

9. ACKNOWLEDGEMENTS

Thanks go out to the interviewees who have volunteered some of their precious and limited time. Your willingness to contribute is very much appreciated and is leading the way into this exciting new field of the maritime industry. Christian Hovden from USN (University of South-Eastern Norway) most of all, who opened the door for me to this fascinating world of autonomy.

10. REFERENCES

1. IMO, 'OUTCOME OF THE REGULATORY SCOPING EXERCISE FOR THE USE OF MARITIME AUTONOMOUS SURFACE SHIPS (MASS)', *Maritime Safety Committee, MSC.1/Circ.1638*, 2021.

2. PLETSCHKE, A., 'Maritime Autonomy and Diversity. Exploring the Role of Technology and Working Practices.', *Thesis for a Master of Engineering (UAS) degree, Novia University of Applied Sciences*, 2024.

3. EMSA, 'Seafarer's Statistics in the EU', *Editions 2020-2022*, https://www.emsa.europa.eu/, 2024.

4. EQUASIS, 'The World Merchant Fleet Statistics from Equasis', *Editions 2020-2022*, https://www.equasis.org/EquasisWeb/public/PublicStatistic?fs=HomePage, 2024.

5. BUREAU VERITAS MARINE & OFFSHORE, 'Seafar NV Receives Approval in Principle from Bureau Veritas for Remote Vessel Operations Technology', https://marine-offshore.bureauveritas.com/newsroom/seafar-nv-receives-approval-principle-bureau-veritas-remote-vessel-operations-technology, 2024.

6. MØRKRID O E., BELLINGMO, P R., and WILLE, E., 'Feasibility Study for an Unmanned Deep Sea Bulk Ship and Short Sea Container Ship', *SINTEF, Trondheim, Norway, Project no. 302005882*, 2023.

7. MALTERUD, K., 'Systematic text condensation: A strategy for qualitative analysis', *Scandinavian Journal of Public Health* 40(8), pp 795-805, doi:10.1177/1403494812465030, 2012.

8. RAYMOND, 'Autonomous ships at a glance with Massterly', https://massworld.news/autonomous-ships-at-a-glance-with-massterly/, 2024.

9. LAZAROWSKA, A., 'A Comparative Analysis of Computational Intelligence Methods for Autonomous Navigation of Smart Ships', *Electronics 13(7)*, https://doi.org/10.3390/electronics13071370, 2024.

10. PEETERS, G., et al., 'An Inland Shore Control Centre for Monitoring or Controlling Unmanned Inland Cargo Vessels.', *Journal of Marine Science and Engineering https://doi.org/10.3390/jmse8100758*, 2020.

11. REACH SUBSEA, 'Reach Remote - A major milestone for uncrewed over-the-horizon maritime operations' *Reach Subsea*, https://reachsubsea.no/wp-content/uploads/2024/03/REA24-2345-100-Brochure-Reach-Remote-digital.pdf, 2024.

12. NEPAL, S., and GUDMESTAD, O T., 'Alarm handling onboard vessels operating in DP mode.' *TransNav Journal, Vol 16*, No 1, pp 51 – 56, DOI: 10.12716/1001.16.01, 2022.

13. United Nations, 'United Nations Convention on the Law of the Sea' UN General Assembly, 1982.

14. MARSHALL ISLANDS MASS CORRESPONDENCE GROUP CHAIR, 'Development Of A Goal-Based Instrument For Maritime Autonomous Surface Ships (Mass)', *MSC 107 Annex 1* (p.52), 2023.

15. ZELIM, https://www.zelim.com/products/unmanned-rescue-vessel/, 2024.

16. JAPAN AND UNITED STATES, 'Development Of A Goal-Based Instrument For Maritime Autonomous Surface Ships (Mass) - Comments on document MSC 108/4/1 regarding network governance' *Maritime Safety Committee 108th session*, 2024.

17. LIBERIA, REPUBLIC OF KOREA AND UNITED ARAB EMIRATES, 'Development Of A Goal-Based Instrument For Maritime Autonomous Surface Ships (Mass) - Network governance', *Maritime Safety Committee 108th session*, 2024.

18. EMSA, 'Seafarer's Statistics in the EU', Edition 2022, https://www.emsa.europa.eu/, 2024.

19. CAI, A., et al., 'Intersectional Analysis of the Challenges and Opportunities of Equitable Remote Operation in the UK Maritime Sector', *Conference: TAS '23: First International Symposium on Trustworthy Autonomous Systems*, https://doi.org/10.1145/3597512.3600202, 2023.

20. JUNG, T., et al., 'Identification of Competences for MASS Operators in Remote Operation Centres', *European Maritime Safety Agency*, 2023.

21. TAE-EUN, K., 'Bridging Education and Future Work: Reframing Education and Competence Model for Remote Maritime Operations (REFRAME)', *The Research Council of Norway*, 2024.

11. AUTHORS BIOGRAPHY

Clemens Plawenn-Salvini is a lecturer at the University of Applied Sciences Emden-Leer. He has sailed in the offshore industry and worked in a crewing company as a training officer. He is educating future nautical officers and is aiming to steadily adapt the education to the newest developments in technology and shipboard operations.

Ove Tobias Gudmestad is a retired professor of Marine Technology (The University of Stavanger), Cold Climate Engineering (UiT, The Arctic University of Norway), and Maritime Technology (Western Norway University of Applied Science). His research activities focus on safety in the design of systems and the operations of technology.

HOW UR E26 AND E27 COULD AND SHOULD APPLY TO DRONES AND AUTONOMOUS VESSELS?

C Soyer, Bureau Veritas Marine & Offshore, France

SUMMARY

This paper will explore the intrinsic reasons for applying UR E26 and UR E27 standards, by digging into existing vulnerabilities, and emphasizing the need for a standardized and interoperable cybersecurity framework. While acknowledging the limits of these regulations, it will address potential challenges and gaps, offering recommendations for mitigation.

Additionally, alternative cyber maritime regulations may be considered to supplement and enhance the overall cybersecurity strategy. In summary, by addressing challenges, fostering collaboration, and considering additional regulatory frameworks, the maritime industry can continue to evolve and safeguard its innovations..

1. INTRODUCTION

The era of autonomous ships is in full swing, marked by rapid technological advances and the growing adoption of autonomous maritime solutions. These vessels, designed to navigate without direct human intervention, promise to improve operational efficiency, reduce costs and increase safety at sea. However, this maritime revolution poses significant regulatory challenges. The need to regulate these emerging technologies and the cyber risk associated to ensure their safe and efficient integration into the international maritime system is crucial.

In this context, resolutions UR E26 and UR E27 of the International Association of Classification Societies (IACS) are essential. Mandatory for all ships signed up for construction after 1 July 2024, they aim to provide a framework for assessing the cyber resilience of ships (UR E26) and onboard systems and equipment (UR E27).

The purpose of this paper is to examine in depth the need to apply the UR E26 and UR E27 regulations to autonomous vessels. In the first part, we will present the objectives and specificities of UR E26 and UR E27. Then, we will discuss the reasons why their application is imperative for the safe and efficient development of autonomous ships. Finally, we will analyze the limitations and challenges posed by these regulations, considering potential improvements to better meet the needs of this revolutionary technology.

2. THE IACS UNIFIED REQUIREMENTS E26 AND E27

2.1 CYBERSECURITY MARITIME REGULATION HISTORY AND CONTEXT

Cyber security emerged in the maritime industry in the early 2000s. The digitalization and automation of on-board equipment was a response to the need for efficiency, competitiveness, and cost reduction. Standardized use of systems such as ECDIS (Electronic Chart Display and Information System) and AIS (Automatic Identification System) is making navigation safer and more accurate. But as with all technological advances, the arrival of this new type of equipment has introduced a multitude of risks, including computer hacking, malicious software, and theft of sensitive data.

Over the next 15 years, the maritime world faced incidents that accentuated the need for stricter frameworks to ensure the cyber resilience of organizations. In June 2017, Maersk, one of the world's largest shipping companies, was the target of a ransomware attack named "NotPetya". After an infected update to accounting software in shore offices, the malware spread across the company's global network crippling all IT data. As a result, 76 ports managed by a Maersk subsidiary were affected, causing a significant financial loss of around 300 million dollars. This incident was a powerful example of the cyber vulnerabilities affecting the entire industry.

Shortly before this event, in 2016, the IMO (International Maritime Organization) published a list of guidelines on maritime cyber risk management, designed to help shipping companies understand and prevent the cyber security risks associated with their operations. But it was in 2017 that it published Resolution MSC.428(98), effective from 1 January 2021, to officially incorporate cyber risk management as an inherent part of security risk management processes. This was an important step towards formalizing cyber security requirements in the maritime industry.

From the point of view of the class societies, it was late in 2022 that the formalization of requirements became official with the publication of Unified Requirement E26 and E27, respectively designed to ensure the cyber resilience of boats and equipment. These two URs have become mandatory for all ships contracted after 1 July 2024.

2.2 UR E26: CYBER RESILIENCE OF SHIPS

The E26 Unified Requirements are aiming to ensure a level of cyber resilience of ships. Applicable to certain types of vessels, from passenger ships engaged in international voyages to cargo ships over 500GT, the requirements introduce a set of controls that can be used to cover a wide range of vulnerabilities. The increasing digitalization of maritime operations calls for rigorous management of onboard systems, requiring a precise and regularly updated inventory to identify the interdependencies between the various systems. To minimize the potential attack surface, requirements focus on the segmentation of networks, all of which must be equipped with specific capabilities to protect against the propagation of malicious agents. This protection requires rigorous management of software, access controls and network monitoring to rapidly identify any abnormal or malicious activity.

As protection is necessary but not sufficient, IACS has also created requirements around operation resumption and continuity, including the need for detailed incident response plans and incident recovery procedures with clearly defined strategies for rapidly restoring capabilities and services critical to ship operations. All these requirements are intended to be applied by all the stakeholders involved in the design, construction and operation of ships, from the shipyards to the shipowners, via the systems integrators and suppliers (who are more specifically concerned by UR E27 detailed below).

2.3 UR E27: CYBER RESILIENCE OF ON-BOARD SYSTEMS AND EQUIPMENT

The E27 Unified Requirements have been created to increase the resilience of onboard systems and equipment in terms of cyber security. They concern both operational technologies (OT), which encompass systems or equipment used to monitor or control physical processes such as propulsion, electrical power generation or navigation systems; and any communication interface based on an Internet protocol, which concern information technologies (IT) such as administrative networks, those intended for passengers, or any other system connected to OT systems on a permanent or temporary basis. It should be noted that UR E27 is the counterpart of UR E26 for equipment, which explains certain similarities in the technical requirements.

The requirements focus on securing systems through user authentication, rigorous management of accounts and identifiers, and protection against malicious software. Communications security is paramount, with measures to guarantee the integrity and confidentiality of data, through the encryption of wireless networks. UR E27 also ensures that the system can continue to operate in the event of a major event, through the ability to operate in degraded mode in the event of a denial of service (DoS) attack, secure management of audit logs, and backups and the ability to reconstruct them.

3. MASS ARCHITECTURE AND CONNECTIVITY CONTEXT

Autonomous marine vehicles are designed to operate without direct human intervention, relying on a complex architecture and interconnection of advanced modules. However, the available research does not agree with the specific systems to be included in these vessels. Based on several studies and the experience gained through the HAZIDs carried out for Bureau Veritas, I have drawn up a list of essential systems.

3.1 AUTONOMOUS SHIPS MAIN COMPONENTS

3.1 (a) Advanced Sensor Module (ASM)

The Advanced Sensor Module (ASM) uses real-time data to navigate, avoid obstacles and analyze the autonomous vessel's environment. It relies on a set of advanced sensors to provide the information needed for decision-making and automatic control. The system includes several typical components, such as radar, lidar, optical and infrared surveillance cameras.

3.1 (b) Autonomous Ship Controller (ASC)

The Autonomous Ship Controller (ASC) supervises and ensures that the necessary decisions are made to navigate in complete safety. Acting like a virtual captain, this central decision-making system uses artificial intelligence models to analyze the environment, strategically plan routes and execute the necessary maneuvers. It is also responsible for communications with the shore-based or remote-control center. It includes several modules necessary for the ship's activity, such as the Anchoring and Mooring System (AMS), the Cargo Handling System (CHS) and the Autonomous Engine Monitoring and Control System (AEMCS), which itself integrates automated sub-systems such as the Autonomous
Navigation System (ANS).

3.1 (c) Autonomous Navigation System (ANS)

The Autonomous Navigation System (ANS) is a set of technologies integrated into the Autonomous Ship Controller (ASC) designed to enable a vessel's navigation to be managed without the need for constant human supervision. Key elements and specialized functionalities of the system include navigation situational awareness through the Navigation and Situational Awareness System (NSAS), collision avoidance through the Collision Avoidance System (CAS), route and speed optimization planning through the Route and Speed Optimization and Planning System (RSOPS), and weather interpretation through the Weather Monitoring and Interpretation System (WMIS).

3.2 CONNECTIVITY CONTEXT AND GLOBAL ARCHITECTURE

3.2 (a) Shore Control Center (SCC)

The Shore Control Centre (SCC) is a land station dedicated to the remote monitoring and control of autonomous vessels. It is important to note, however, that it can be deported to a manned vessel. Equipped with advanced automation technologies and decision support systems, the SCC centralizes all the functions needed to manage autonomous vessels from land. This includes a remote bridge and engine control modules, enabling direct control of the vessels. Thanks to this infrastructure, qualified personnel can perform critical tasks such as coordinating emergency response, assisting navigation and ensuring compliance with regulations, without having to be physically present on board the ships.

3.2 (b) Communication System

Communication systems are essential to provide the connectivity needed for surveillance, control and coordination between autonomous vessels and the SCC. These systems include AIS, VHF radio, VDES (VHF Data Exchange System) and GMDSS (Global Maritime Distress and Safety System). Thanks to these technologies, autonomous vessels can improve their situational awareness and exchange crucial information with other vessels at sea as well as with shore-based facilities.

To ensure effective communications, 2 categories of technologies are used:

- LoS (Line of Sight):
  - Land-based communications - VHF (Very High Frequency), Broadband and Internet generally require a direct line of sight between communications antennas. However, in the maritime context, Broadband and Internet communications are sometimes transmitted beyond the line of sight using relays, extended networks and long-distance links.
- BeLoS (Beyond Line of Sight):
  - Satellite Communication System (SATCOM) enables transmissions beyond the line of sight, using orbiting satellites to relay signals between distant points. A key and indispensable technology for autonomous ship operations, using robust communication channels to ensure reliable and consistent exchanges.

3.3 GLOBAL MASS ARCHITECTURE DIAGRAM

Here is an overview of the global architecture of an autonomous vessel based on the equipment previously presented:

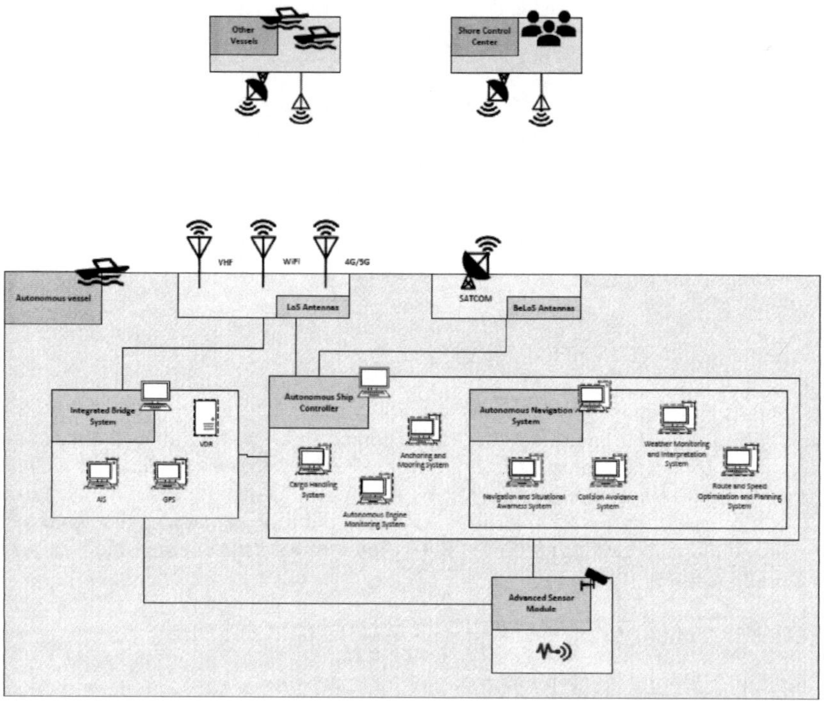

Figure 1: Global MASS Architecture diagram.

4. APPLICATION OF UR E26 AND UR E27 TO AUTONOMOUS VESSELS

The previous section has enabled us to understand that autonomous ships have a complex architecture and a dependence on advanced communications systems. This section will now explore the application of UR through the prism of organizational, technical and documentation aspects.

4.1 ORGANIZATIONAL ASPECTS

4.1 (a) Risk management and business continuity

UR E26 includes proactive risk management to identify and mitigate potential threats to autonomous vessels. Systems within the scope of the UR must have well established business continuity and recovery plans/ They must include strategies to isolate compromised systems from the rest of the ship and quickly restore critical functions. Artificial intelligence models already integrated into systems such as the ASC can improve anomaly detection and enable rapid response to incidents.

4.1 (b) Coordination and communication

To manage operations and cyber incidents effectively, coordination with the SCC is fundamental. Despite the need of details for certain issues, such as secure management of telecommunication systems, URs stress the importance of securing every IP based system including navigation and communication ones. A well-defined organization is needed to manage communication systems in the event of disruption, with continuity plans in place to ensure resilience of operations. By focusing on these organizational aspects, operators can enhance the safety and efficiency of autonomous maritime operations, while complying with regulatory requirements.

4.1 (c) Training and awareness-raising

In accordance with UR E26, SCC staff and operators of autonomous vessels must receive in-depth training on best cybersecurity practices, covering both an introduction to the subject and the best practices to be put in place against human attacks (phishing, identity fraud), but also covering security protocols, incident management and response procedures. This training can consist of theoretical learning sessions, as well as regular simulation exercises that enable teams to reinforce their ability to act quickly.

4.2 TECHNICAL ASPECTS

4.2 (a) Security of critical systems

Essential systems, such as ASM and ANS, are at the heart of the functionality of autonomous ships. These systems enable navigation, obstacle avoidance and autonomous decision-making. In line with the requirements of UR E27, it is imperative that these systems are protected by robust security measures to defend sensitive data and communications against cyber-attacks. System access management is one of the crucial aspects, with multi-factor authentication and role-based access control (RBAC) ensuring that only authorized users can interact with critical systems.

4.2 (b) Securing communications

In the context of autonomous navigation, exchanges between the SCC and the ship are crucial. Their security can be ensured by the UR E27, which requires the use of encryption technologies such as TLS (Transport Layer Security) and IPSec (Internet Protocol Security) to guarantee the integrity and confidentiality of the information transmitted. This encryption ensures that important information, such as navigation commands, remains intact and secure. The UR E27 also requires the implementation of robust authentication mechanisms, such as multi-factor authentication (MFA), to ensure that only authorized entities can enter the ship's network. The UR E26 places the emphasis on continuous monitoring of communications to detect unauthorized access attempts or any other anomalies. All these requirements would therefore guarantee the reliability of the ship's navigation and control systems.

4.2 (c) Complete defense against cyber threats

A complete defense is effective when it covers all the stages of an attack in chronological order. UR E26 is inspired by NIST, which proposes a structured approach to managing cyber threats. The five main functions are: Identify, Protect, Detect, Respond and Recover. The technical requirements fall under several of these functions.

The URs underline the importance of implementing protection systems, such as installing firewalls to filter network traffic and create barriers against unauthorized access. Threat detection can be based on different systems, or on the use of intrusion detection and prevention systems (IDS/IPS), which monitor network activity and report anomalies. In response to detected incidents, protocols must be put in place to isolate compromised systems and coordinate actions with the SCC. The recovery phase involves regular backups of data and system configurations, as well as recovery plans to quickly restore critical functions and minimize interruptions, ensuring continuity of operations and the safety of navigation and control systems on autonomous vessels.

4.3 DOCUMENTARY ASPECTS

4.3 (a) Inventory of Computer Based Systems (CBS)

The identification part presented by the NIST allows the effective management of digital assets, whether at equipment level with UR E27 or at ship level with UR E26, by keeping a detailed inventory of the CBS, which must include all the hardware and software components used. It must specify a significant amount of information such as the version of the operating systems or the network interfaces present and used. If we take the example of the ASC, the system must be fully documented, including the modules and sensors connected and the communication protocols used.

This exhaustive documentation not only enables digital assets to be managed effectively, but also enables the security updates and patches needed to keep systems up to date with new threats to be planned. By applying the requirements throughout all stages of a vessel's life, UR E26 recommends regular audits of this inventory to ensure that all systems are still compliant with established security standards and that no components are obsolete or exposed to unnecessary risk.

4.3 (b) Documentation of security zone topology

Mapping network connections and security zones is key to identifying potential weaknesses and planning the necessary countermeasures. It enables security teams to identify points of vulnerability where cyber-attacks could occur, as well as possible paths for threats to spread. UR E26 highlights network segmentation as an essential security measure. This segmentation makes it possible to isolate different security zones, restricting the lateral movements of potential attackers in the event of a system being compromised. For example, in the event of an intrusion into a less secure segment, such as the network used for routine communications or interactions with management systems, network segmentation prevents this threat from spreading to critical autonomous navigation or propulsion control systems.

4.3 (c) Secure Development Life Cycle (SDLC)

A Secure Development Life Cycle (SDLC) is fundamental to ensuring that embedded software and IT systems are designed, developed and maintained to the highest safety standards. Patch management is an integral part of the SDLC for autonomous vessels. It involves the regular updating of software and operating systems to correct identified vulnerabilities. This is particularly important for critical systems such as the ASC, where vulnerabilities could be exploited to disrupt ship operations. Security testing must be rigorous and include penetration assessments and robustness tests to ensure that systems are resilient against cyber-attacks.

5. THE LIMITS OF CURRENT REGULATIONS

For traditional, non-autonomous ships, one of the final common safeguards against incidents is the presence and intervention of human crew members. They can monitor the vessel's systems, identify and respond to potential issues, and make real-time decisions to ensure the safety of the ship, its cargo, and its passengers. However, with the advent of autonomous ships, this human intervention safeguard is replaced by communication capabilities and algorithmic decision-making. Given the potential absence of direct human oversight, it is crucial to establish and apply robust norms and standards to compensate for this change.

5.1 SENSOR INTEGRITY AND SAFETY

Autonomous vessels rely on a variety of sensors and navigation systems such as GNSS, AIS, radar and cameras to collect data essential to their safety and operational efficiency. These systems are vulnerable to attacks such as spoofing, jamming or the injection of falsified data. For example, falsified GNSS signals can mislead navigation systems, and manipulated AIS data can compromise ship safety. Similarly, radar and camera signals can be spoofed or altered, making it difficult to detect obstacles. Although URs prescribes measures for intrusion detection and recommends the use of firewalls and network segmentation, these standards lack precise guidelines for guaranteeing data integrity.

A major recommendation for enhancing cyber resilience is to ensure redundancy of critical systems, for example by using separate antennas for AIS and GPS/GNSS signals to secure the information received and sent, or by coupling the GPS/GNSS signal with a gyrocompass linked to systems such as the ANS to compensate for any loss of signal. These measures, complemented by signal filtering technologies and anomaly detection algorithms, are essential to ensure that navigation decisions and operations are based on reliable and secure data.

5.2 PROTECTION OF ARTIFICIAL INTELLIGENCE SYSTEMS

Machine learning systems are central to the operation of autonomous ships. Artificial intelligence is used to make decisions, control actions and detect situations. These systems can therefore be the target of data poisoning attacks where biased data disrupts the algorithms. The URs include requirements for protection against the use of malicious code, multifactor authentication and intrusion detection systems, but vulnerabilities persist. For example, training data validation processes and the integration of anomaly detection mechanisms specific to AI models are not explicitly covered. To strengthen cyber resilience, it is crucial to improve safety practices, including strict validation of the data used to learn models and regular auditing of AI systems to detect and respond to anomalies.

5.3 THE SHORE CONTROL CENTER (SCC)

To guarantee the safety of autonomous ships, it is essential to include shore control centers in the regulations, especially when they use outsourced infrastructures in the cloud or virtual machines. Existing regulations URE26 and URE27 focus primarily on the security of systems on board ships, often neglecting critical infrastructures located remotely. Yet these control centers, which supervise and control ships, are often hosted in cloud or virtual environments, and play a central role in the operations of autonomous ships. For complete cyber protection, it is crucial to extend regulations to these shore-based infrastructures, but the scope of the study should be clearly identified so as not to become an analysis of the whole of IT. This includes implementing security measures such as remote or on shore communications encryption, network segmentation and robust protocols for access management. In addition, it is recommended that redundant architectures be put in place, such as backup servers and recovery plans, to ensure continuity of services in the event of a cyber-attack. By fully integrating shore-based control centers into the regulatory framework, the protection of the entire autonomous ship ecosystem against cyber threats can be strengthened, ensuring the security and resilience of maritime operations.

6. CONCLUSION

A detailed analysis of the UR E26 and UR E27 regulations highlights their crucial importance in the framework of autonomous ships. These regulations, developed by the International Association of Classification Societies (IACS), aim to guarantee the safety and cyber resilience of ships in the face of the growing risks associated with digitalization. UR E26, which focuses on the cyber resilience of ships, and UR E27, which specifically targets onboard systems and equipment, are essential to protect autonomous ships against cyber threats. These regulations cover measures such as multifactor authentication, network segmentation and continuous monitoring of systems, enabling effective prevention and detection of intrusions.

However, despite their usefulness, these regulations have certain limitations. The integrity and security of critical sensors such as GNSS and AIS remain vulnerable to attacks such as spoofing and jamming. In addition, artificial intelligence (AI) systems require more rigorous data validation protocols to prevent injection attacks. In addition, land-based infrastructures, which are often outsourced, do not yet benefit from sufficient regulatory coverage, leaving an opening for potential cyber-attacks on shore-based control centers (SCCs).

To overcome these limitations, additional regulations may be needed. The International Maritime Organization (IMO) is currently working on a draft code for MASS, which will include concepts covering the use of artificial intelligence in systems and the management of remote operations, but it will not be published before 2025 and applicable before 2027. There are no regulations yet that will cover all aspects at once. However, working like a puzzle, a few guidelines could be put together to open the scope, such as the IMO "Guidelines on maritime cyber risk management" already mentioned, but also DNV guideline 0264 "Autonomous and remotely operated ships". Within Bureau Veritas Marine Offshore, we are currently working on a revision of our NR 659, which already covers the requirements of URs through several notations, to provide an adapted notation for ultra-connected vessels such as autonomous vessels.

8. ACKNOWLEDGEMENTS

I would like to thank Arthur PIQUEE-AUDRAIN, Benoit BOUERY, Panagiotis ANASTASIOU and Philippe VAQUER from the cyber security team as well as Bérénice LE GLOUANEC, Ilia MASLOV and Jérôme FAIVRE from the digital autonomous ship team at Bureau Veritas Marine & Offshore, for their meticulous proofreading of my paper as well as their precise and very useful comments.

9. REFERENCES

1. Guidelines for Cyber Risk Management in Autonomous Shipping, ITRUST, 18th January 2024.

2. CHO, S., ORYE, E., VISKY, G., and PRATES, V., 'Cybersecurity Considerations in Autonomous Ships', CCDCOE, 2022.

3. WANG, J., ZHOU, K., XING, W., LI, H., and YANG, Z., 'Applications, Evolutions, and Challenges of Drones in Maritime Transport', MDPI, 27th October 2023.

4. WENNERSBERG, L A L., NORDAHL, H., RODSETH, O J., 'Autonomous ship design methods and test principles v1.1', AUTOSHIP D3.2, 5th March 2021.

5. Guide de sensibilisation des acteurs du transport maritime & fluvial et des opérations portuaires aux risques de brouillage du GNSS (GPS, Galileo, Glonass, Beidou, etc.), Ministère de la transition écologique et de la cohésion des territoires Agence nationale des fréquences (ANFR), February 2024.

6. YOO, J., and JO, Y., 'Formulating Cybersecurity Requirements for Autonomous Ships Using the SQUARE Methodology', MDPI, 24th May 2024.

7. VEITCH, E., and ALSOS, O A., 'A systematic review of human-AI interaction in autonomous ship systems', NTNU, 12th April 2022.

8. LLOP, M., ARA, P G., LOPEZ, J M L., and MALDONADO, J A G., 'D5.1 – ModalNET Specifications, systems architecture and design of cyber-secure communication', SEAMLESS, 1st March 2023.

9. PAPAGEORGIOU, S., 'MASS, The Evolving Path of Technology', IMO-ROK Symposium, 14th May 2024.

9. AUTHORS BIOGRAPHY

Charlotte SOYER holds the current position of Cyber Security Analyst at Bureau Veritas Marine & Offshore. After 1 year with a Big Four consulting firm, she joined the BV team in September 2023. Organizational and technical architecture audits was her specialty, but she also has a keen interest in network management and systems administration. Today, she is seeking to specialise in studying the cyber security regulation of hyper-connected vessels such as drones and autonomous ships, as well as training in electronic warfare techniques.

SAFETY ASSURANCE OF AUTONOMOUS SURFACE VESSELS

E Rajabally, M Wylie, and **E Turkbeyler**, BMT, UK

SUMMARY

Safety assurance in the case of autonomous sea going vessels is nontrivial due to the pace of change in enabling technologies and their disruptive impact. Leaning on proven practice, management of safety risk to a tolerable level and subsequent demonstration of a safety case remain pivotal to safety assurance of maritime autonomy. The long-established discipline of functional safety engineering concerns mitigating the risks of system or component failures that would otherwise cause harm and numerous supporting analysis techniques abound. However, in the case of autonomous systems, risk assessment must go much further and address disruptive technologies such as Artificial Intelligence (AI). Despite this, detailed thinking on safety practice tuned to maritime autonomy is emerging and good practice may also be read across from analogous domains such as self-driving road vehicles. The authors present a summary status of safety assurance in the context of maritime autonomy.

1. INTRODUCTION

Some degree of autonomous operation (i.e. without human monitoring and control) has featured onboard numerous small uncrewed underwater vehicles and small surface vessels in routine use, and there is appetite worldwide to scale its adoption to larger and more complex vessels.

In the naval domain, the US openly discuss their medium displacement uncrewed concept, a 45-190ft patrol sized vessel mooted for Intelligence, Surveillance and Reconnaissance (ISR) and Electronic Warfare (EW) and their large displacement concept, a 200-300ft corvette sized vessel mooted for Anti-Surface Warfare (ASuW). In the commercial domain meanwhile, the 262ft Yara Birkeland container ship has been transitioning to autonomous shortsea voyages in Norway since 2022 whilst Japan is keenly pursuing their MEGURI2040 vison for autonomous operation of cargo ships in the round of up to 650ft.

Given the ambition, the safety of autonomy at sea has been attracting increased attention and this paper provides summary status of the topic. The next section presents the case for focussed attention on the topic and this is followed in the third section, by a survey of the state of practice concerning safety of autonomy at sea. As a pivotal and growing enabler to autonomy, the safety of AI is considered in the fourth section and conclusions follow in the fifth.

2. WHY WORRY ABOUT SAFETY OF AUTONOMY

Many advocate for and extol the benefits of adoption of autonomy at sea such as increased safety. The tantalising prospects are to both reduce the scope for human error during safety-critical decision making and in de-crewing, to completely remove people from danger. A recent demonstration of the benefit from the latter is the failed touch-down of a Falcon 9 uncrewed reusable booster rocket in August (flight 367) on a SpaceX barge in the Atlantic Ocean. The rocket toppled over in flames on to the landing deck but with the barge itself self-navigating and uncrewed, no personnel were endangered. With any engineering initiative, assurance of safety and so prevention of harm to people, property and the environment remains paramount. Until the body of evidence from operations at sea indicates a reduction in unsafe events, it is sensible therefore to probe the hypothesised benefits.

The European Maritime Safety Agency (EMSA) tracks casualties from incidents at sea recorded by EU member states and a summary of fatalities and injuries between 2014 and 2022 is given in Table 1. Supporting the benefit of de-crewing and so the case for autonomy adoption, the prospect is to dramatically reduce onboard accidents and so "person occurrences". Summing up figures in the fourth and sixth columns, 36% of fatalities were due to such onboard "person occurrences" and a staggering 79% of injuries. That said, the particular EMSA statistics in Table 1 express neither likelihood of the accident nor consequence severity, the pillars of safety risk. Taking "ship occurrences" from Table 1 for example, as we cannot evidence with any statistical merit that a collision is more or less likely for conventional as compared to an autonomous vessel then accident occurrences and the subsequent fatality and injury toll are unknown. If we simply assume the same underlying number of ship occurrences in moving from the former to the latter, then whilst the fatality and injury count might be expected to fall (from the numbers reported below) with a reduction or removal of onboard crew, we cannot assert to what degree.

Collision, the single biggest "ship occurrence" contributor to fatalities of 182 or 20% of the total, is clearly between two or more vessels and the EMSA figures do not distinguish between them. So, collision between uncrewed, autonomous vessel and conventional vessel with people onboard, might have no fatality or injury implication to the former but unlikely so for the latter. Looking to "person occurrences", we might assume autonomous vessels with no or at least fewer crew, to suffer fewer fatalities and injuries such as due to onboard slips and trips. However, we should again resist assuming the degree of reduction with crew needing to board for example for maintenance and cargo operations. All considered, anticipating the implications of autonomy to accidents at sea is not trivial and we should resist second guessing. Crucially, incident and accident statistics disguise the countless hours of operation that pass by without unsafe event due to vigilance and expert decision-making of skilled seafarers. Human ability to avert danger goes unrecognised and unquantified in such statistics.

Table 1. Shipping Fatalities and Injuries in EU Waters, 2014-22 taken from [1, pp. 30-33].

Accident or Incident	**EMSA category**	**Fatalities**	**% of total**	**% of total**	**Injuries**	**% of total**	**% of total**
Collision	Ship occurrences	182	20%	63%	377	5%	21%
Flooding/foundering		128	14%		84	1%	
Capsizing/listing		103	11%		82	1%	
Other		152	17%		1017	14%	
Slipping, stumbling, falling	Person occurrences	163	18%	37%	1776	24%	79%
Body movement without stress		40	4%		1238	17%	
Deviation by overflow, overturn, leak, flow, vaporisation, emission		34	4%		309	4%	
Other		94	10%		2462	34%	
	Totals	896			7345		

2.1 A NEW BREED OF ACCIDENT EVENT SEQUENCES

The safety bowtie is a well-established visualisation of the cause-effect chain from multiple potential initiating events through to potential accidents via intervening hazard, the bowtie 'knot'. It has been applied to safety analysis of autonomous vessels and an example is shown below (Figure 1) to help illustrate a concern that autonomy raises. The right-hand accident and emergency situations and the instigating hazards are familiar and apply both to autonomous and conventional vessels. However, there are different and unfamiliar instigating events on the left-hand side such as "Autonomous control failure" and "Hostile remote control" which will demand potentially untried and untested mitigations or lines of defence both upstream and downstream of the hazard that sever the lines joining the knot to either bow. A new breed of accident event sequences resulting from the adoption of autonomy at sea presents significant challenge to safety assurance.

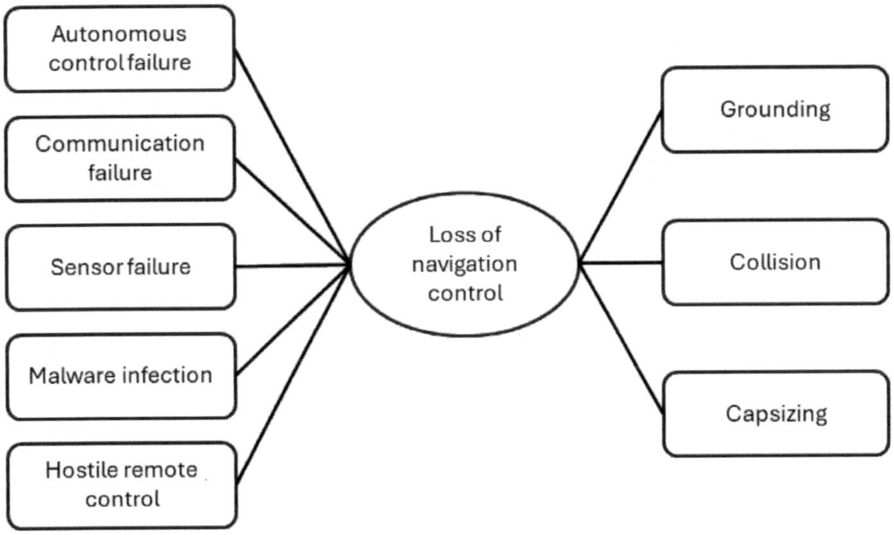

Figure 1: Partial safety bow tie example for an autonomous vessel [adapted from reference 2].

2.2 THE POTENTIAL PERIL OF HUMAN AUTONOMY TEAMING

The authors of this paper have published elsewhere on the specific nuances of autonomous vessels that make their safety assurance more difficult than for conventionally crewed vessels [3]. One such nuance explored further here is a blurring of responsibility between human and machine. For an autonomous vessel of any complexity, the human must act in concert or team with the machine in a balance that varies firstly according to the system (e.g. navigation versus say cargo management) and its operational phase (e.g. open water transit compared to harbour approach). Secondly, this balance will also likely shift according to changing conditions of both the system in question (e.g. according to its health) and its environment (e.g. varying traffic congestion). Historically, there has been clear and fixed distinction in role allowing for largely independent approaches to demonstrating that a vessel is 'safe to operate' from an equipment perspective as opposed to demonstrating that it 'is operated safely' by its crew. The potential peril of a blurring of responsibility between human and machine brought about by increased autonomy adoption is illustrated as follows by a well-publicised road accident fatality.

In March 2018, an Uber self-driving test car struck and killed Elaine Herzberg as she wheeled a bicycle across the road in Tempe, Arizona. The US National Transportation Safety Board determined that the probable cause of the crash was "the failure of the vehicle operator" (Rafaela Vasquez) "to monitor the driving environment and the operation of the automated driving system because she was visually distracted throughout the trip by her personal cell phone" [4]. On approaching Elaine Herzberg, the automated driving system repeatedly misclassified her as a vehicle, a bike and an unknown object triggering different conflicting actions. Upon realising impact was unavoidable, control was handed back to its human safety driver, Rafaela Vasquez who was tasked with actively monitoring the self-driving technology and the road ahead. Vasquez began braking only after the car had already hit Herzberg who was thrown 23 metres by the impact and died of her injuries at the scene. Dash-cam footage released by police showed Rafaela Vasquez looking down, away from the road, for several seconds immediately before the crash. Originally charged with negligent homicide, Rafaela Vasquez pleaded guilty to a reduced charge of endangerment in 2023 and so avoided a prison sentence. Flagged by various sources as a case of 'automation complacency', the Uber accident is a striking example of potential severe danger in human autonomy teaming.

2.3 REGULATORY SHORTFALL

Conventional assurance of safe operation by a crew follows from personnel training and qualification that ultimately derives from regulation such as the IMO instruments. With the blurring of operational responsibility between human and machine, the question then raised is whether such instruments and their derivations provide sufficient safety regulations for those roles that machines take over from humans. Autonomous navigation has been a fertile development area for some time and here the 'rules of the road', the IMO's

International Regulations for Preventing Collisions at Sea (or COLREGs for short) have primacy but were written exclusively for human consumption.

In isolation, the COLREGs are insufficient instructions for machine execution and a few examples follow to illustrate this. Firstly, Rule 2 considers responsibility for a vessel in collision encounters and permits departure from the Rules to avoid "immediate danger". Whilst experienced seafarers would know to recognise such a situation at sea for example from assessing time and distance to the closest point of approach, the lack of specific COLREGs description of immediate danger is insufficient for machine execution. Secondly, the COLREGs cite innate human traits and sections 2 and 3 for instance, specify navigational conduct in visual sight and in restricted visibility respectively. Finally, the COLREGs refer to conditions for actions or behaviours that are unquantified. Rule 8 for example, specifies actions to avoid collisions and requires that any alteration to course or speed be large enough to be "readily apparent" to another vessel without going on to quantify this beyond avoiding "a succession of small alterations of course and/or speed". Whilst COLREGs have served perfectly well for human compliance and are pivotal to navigational safety of conventional vessels, stand-alone they fall short for machine execution and autonomous operation.

3. EMERGING AUTONOMY SAFETY GUIDANCE

In light of the specific safety pitfalls of autonomy adoption at sea, a few of which are drawn out in the previous section, there is considerable development of relevant safety practice as summarised in the following subsections.

3.1 CONTINENTAL EUROPE LEADS ON MARITIME SAFETY ANALYSIS

Undertaken by the consultancy arm of Norwegian classification society Det Norske Veritas (DNV), the EMSA has commissioned development of a Risk Based Assessment Tool (RBAT) for the autonomous and remote control of sea going vessels. Four reports are available with the fifth and final still outstanding. The fourth report explains the current status of RBAT and describes its application to three hypothetical case study vessels with varying onboard autonomy: a short-sea cargo vessel, small passenger ferry and Ro-Pax ferry [5].

RBAT Part 1 is the detailed description of the use of autonomous and remote control by subdivision of a vessel's "mission" into phases, operations, control functions and control actions. RBAT Part 2 is then a hazard analysis of all control actions. Unsafe conditions are first identified by considering a provided list of control action deviations from design intent, similar to conventional HAZard and OPerability study (HAZOP) "guidewords", such as the control action occurring too early. Part 3 of RBAT is a mitigation analysis of Part 2's unsafe conditions. This entails understanding any Fault Detection, Isolation and Recovery (FDIR), particularly identifying independent prevention and mitigation measures and ranking their effectiveness. RBAT Part 4 is a risk matrix evaluation and departs from the convention of considering event likelihood (given the difficulty in quantifying this) and consequence severity to instead consider consequence severity (from RBAT Part 2) and mitigation effectiveness (from RBAT Part 3). Each unsafe condition is thus ultimately categorised as high risk (intolerable), medium risk (tolerable) or low risk (As Low As Reasonably Practical, ALARP). RBAT Part 5 focuses on driving risk down to tolerable and ALARP by for example, increased control action integrity, introducing operational restrictions, and better mitigating the unsafe condition and the hazard severity.

Bolbot and Theotokatos present the safety approach developed by EU Horizon 2020 collaborative project AUTOSHIP and its application to hypothetical cases studies of remote and autonomous operation of a short sea shipping cargo vessel (i.e. undertaking short coastal journeys) and an inland waterways barge [6]. They compare eleven different risk and hazard analysis methods and ultimately advocate Preliminary Hazard Analysis or HAZard IDentification (HAZID) citing its applicability irrespective of the maturity of design and its support from class societies. The AUTOSHIP safety analysis steps are similar to those endorsed by the EMSA [5] and described above.

3.2 RECOMMENDED SAFETY PRACTICE FROM CLASS SOCIETIES AND OTHERS

Lloyd's Register (LR) was early to develop a design and construction code for Unmanned Marine Systems [7]. The LR code attributes required system "levels of integrity" in light of the consequence of system failure to people safety, namely high (unacceptable), medium (conditionally acceptable) or low (unconditionally acceptable). Alternatively, DNV set out assurance expectations according to four areas of navigation, engineering/platform systems, remote control and communications [8]. As DNV's guidance pertains to both autonomous and remote operation, they are unique in advocating explicit analysis of human factors related risk via dedicated methods beyond the conventional.

Bureau Veritas (BV) and the American Bureau of Shipping (ABS) focus on risk classification. BV proposes potential hazard types associated with various vessel "functional groups" and define risk as an additive combination of indices for causal event frequency and severity of hazard consequence [9]. ABS assigns risk category of high, medium and low according to the consequence of a functional failure (distinguished as not dangerous, potentially dangerous and immediately dangerous) and both whether human supervision is onboard and remote, and whether it continuous, periodic or on-demand [10].

Finally, on behalf of an industry collective, MARITIME UK (2023) describe a voluntary code of practice now in its seventh edition, signed up to by many suppliers and relevant other parties [11]. In terms of safety of autonomy, the code steps back and proposes element of a safety management system to include policy, appropriate responsibilities, culture, risk management, procedures for safe operation and emergency response, personnel and training, along with equipment maintenance.

3.3 SAFETY READ ACROSS FROM ADJACENT DOMAINS

Considerable safety insight can be gleaned from the adoption of autonomy outside of maritime and the industry should strive to avoid reinventing the wheel. Two of many initiatives of note are highlighted below as they potentially adopt a stance to safety assurance that is unfamiliar to the commercial maritime sector.

Goal Structuring Notation (GSN) is a popular means of safety case presentation adopted by other sectors. The UK's University of York have then proposed SACE (Safety Assurance of Autonomous Systems in Complex Environments), a set of safety case GSN "patterns" to structure the safety argument of an autonomous system [12]. The top-level goal or claim is that the autonomous systems will sufficiently mitigate all hazards both within and outside of its defined operating context and a recurring GSN pattern then applies at each level of system decomposition, that safety requirements are met and any potential hazardous failures are managed.

The automotive domain has led on handling hazards in the absence of equipment faults. The ISO/PAS 21448 'Safety of the Intended Functionality' (SOTIF) standard considers the mitigation of risk due to unexpected operating conditions, gaps in requirements and foreseeable misuse [13]. Valuable safety insight is anticipated from the application of SOTIF to maritime autonomy. Clear bounding of a vessel's concept of operation has long been common precursor to safety assessment irrespective of particular method chosen and rigorously identifying foreseeable excursions beyond this boundary is the kick-start to such SOTIF thinking.

4. SAFETY OF ARTIFICIAL INTELLIGENCE (AI)

The Ada Lovelace institute finds that AI safety is not an established term and can be interpreted in many ways. They differentiate four types of harm: accidental (e.g. failures), misuse (e.g. deliberate misinformation), structural (e.g. social, political and economic), and upstream (e.g. inappropriate use of personal data) [14]. In recognition of the growth in interest in AI, the previous UK government established the "AI Safety Institute" in 2023. This was claimed to be the first state-backed organisation in the field and define AI safety as "The understanding, prevention, and mitigation of harms from AI. These harms could be deliberate or accidental; caused to individuals, groups, organisations, nations or globally; and of many types, including but not limited to physical, psychological, social, or economic harms" [15].

Habli and McDermid consider the growth in scope of safety of AI based systems compared to safety of conventional systems and query whether the traditional discipline of safety engineering is equipped to deal with the former and whether it is indeed feasible to do so [16]. They note that in safety-critical industries such as transport, the safety management focus has been on accidental physical harm to humans (e.g. death and injury) and damage to property. A shift to include harm to the natural environment was a subsequent addition and more recent still, psychological damage and deliberate harms (i.e. those inflicted by malicious actors). The latter has blurred the boundary between safety and security disciplines where threats to the latter can snowball into hazards and so have implications for the former. Adoption of AI heightens the challenge due for example, to its proneness to adversarial attack. A potential broadening of safety to encompass moral, political, social and financial harms, whilst unfamiliar to safety practitioners, should not be neglected.

4.1 SAFETY ASSURANCE SCOPE

As guidance is still in development at the time of writing, there is no broad agreement on what constitutes sufficient assurance or compelling case for AI safety. Furthermore, whilst there are common core topics that fall under the scope of AI assurance such as training data, model selection, verification and validation and system implementation, there are satellite topics that are less common. Table 2 lists the range of assurance topics offered from several prominent resources with the delineation of rows used to roughly align similar areas as interpreted by the authors of this paper. Despite there being some gaps in coverage along with potential terminology conflicts, Table 2 illustrates the need to take a 'big picture' and full lifecycle view of AI safety assurance.

Table 2. Comparing key perspectives on the scope of safety assurance for Artificial Intelligence (AI).

The Alan Turing Institute's AI lifecycle, 2024 [17]	**Dstl Biscuit Book Assurance Aspects, 2021 [18]**	**ISO-IEC-23053-2022 ML Pipeline, 2022 [19]**	**AMLAS Assurance Stages, 2021 [20]**	**DSIT AI Assurance Spectrum, 2024 [21]**	**IET 10 Safety Assurance Pillars, 2024 [22]**
Project planning			Assurance scoping		Hazard & risk assessment
Problem formulation	Requirements		Requirements assurance		Specification & architecture
	Computing hardware				
Data extraction / procurement	Data	Data acquisition	Data management	Training data	Data
Data analysis		Data preparation			
Preprocessing					
Model selection & training	Algorithms	Modelling	Model learning	AI models	Machine learning (ML)
Model testing & validation		Verification & validation	Model verification		Verification, validation & assurance
Model reporting					
System implementation	Integration	Model deployment	Model deployment	AI systems	Algorithmic behaviour
	Systems				
	Systems of systems			Broader operational context	Legal & ethical considerations
	Adversaries				Security
User training					Human factors in AI safety
System use & monitoring		Operation			Maintenance & operation
Model updating					

Whilst the reader is directed to the sources referenced in Table 2 for topic specific know-how, there are some generic assurance guidelines that apply. Process-based assurance generates confidence from the way in which something is developed whilst product-based assurance generates confidence in a final outcome, both methods generally benefit the building of a compelling safety case. The former concerns adoption of good processes i.e. those known to consistently deliver a well-engineered solution as well as quality assurance that they have been followed. The latter concerns test, evaluation, verification (building the thing right) and validation (building the right thing). To provide a flavour of specific AI assurance techniques, the UK Department for Science, Innovation and Technology (Dstl) lists key examples as follows [21]:

- Risk assessment
- (Algorithmic) impact assessment
- Bias audit

- Compliance audit demonstrating adherence to internal policies and external regulations
- Conformity assessment that a system meets its performance specification
- Formal verification that a system satisfies specific requirements

4.2 AI SAFETY RISKS AND ENABLERS

AI or otherwise related, safety risks concern the likelihood and consequence severity of hazards (i.e. events with the potential for harm). At the highest level, safety risks whether AI related or otherwise, must be identified, assessed and managed [23]. Different AI specific safety risks as proposed by the Alan Turing Institute are listed in Table 3 and the subsequent high-level failure modes as differentiated by the UK Institution of Engineering and Technology [22] are as follows:

- Model input error, such as a sensor error or an input outside the training data set of the model.
- Incorrect output caused by model bias i.e. undue weighting that is not representative of a more general case.
- Incorrect output caused by lack of algorithm robustness.

Table 3. Artificial Intelligence (AI) specific safety risks [summarised from 17].

Risk Implication	**Risk Type**	**Brief Risk Description**
Performance & Reliability	Concept or Model Drift	Vulnerability to changes in the underlying distribution of data
	Brittleness	Previously 'unseen' inputs that trigger unexpected inferences
	Overfitting	Noise in training dataset is incorrectly interpreted as underlying data distribution
	Unpredictability "in the wild"	Self-updating systems coping with continuously changeable surroundings
Security and Robustness	Adversarial attacks	Malicious falsification of input data to induce misclassification, for example
	Data poisoning	Malicious modification of training, validation and testing data
	Transfer learning attacks	Malicious manipulation of pre-trained models to poison task specific models
	Misdirected Reinforcement Learning	Insufficient control on reward function leading to harmful consequences

The Alan Turing institute cite key enablers to AI safety as performance, reliability, security and robustness [17]. Definitions of these enablers are given and measures of performance characteristics. To pick out a few that are key, they define performance as accuracy (i.e. correctness), precision (i.e. the ratio of true positives to the sum of true positives and false negatives) and recall (i.e. the ratio of true positives to the sum of true positives and false positives). Reliability is adherence to intended functionality or design specification whilst security is protection from unauthorised modification or damage to the AI or access to private data. Finally, robustness is performance under harsh conditions. Usefully, the Alan Turing institute goes on to detail safety considerations or assurance activities related to each of the safety enablers for the different lifecycle stages listed in Table 2.

To conclude this section, brief mention is made of standards given a recent proliferation of development related to AI. Simply put, a standard is agreed best practice. Under the umbrella of trustworthy and responsible AI, big industry IT players such as Google and Microsoft publish their own standards that they adhere to. There are then many multi-party generated standards with national and international backing such as from the International Organization for Standardization (ISO). To illustrate, UKRI (2023) identify circa 16 AI related standards with a focus on safety with almost all aimed at road vehicles [24]. Meanwhile, the AI standards hub (https://aistandardshub.org/) lists 56 standards against the topic safety but these mainly concern automated and autonomous systems in road transport, industrial robotics and healthcare. Finally, the Organisation for Economic Co-operation and Development (OECD) host an online "Catalogue of Tools & Metrics for Trustworthy AI" (https://oecd.ai/en/catalogue/tools) with 174 tools having safety as objective at the time of writing. With so many standards and tools, four general purpose ISO publications are highlighted to the reader:

ISO/IEC 42001:2023 on AI management systems, ISO/IEC TR 24028:2020 on AI trustworthiness guidance, ISO/IEC TR 5469:2024 on functional safety of AI, and ISO/IEC 23894:2023 on AI risk management.

5. CONCLUSIONS

The adoption of autonomy to vessel operation at sea has, until recent times, been applied to cases with little perceived change to safety. Emerging and future cases with greater ambition promise great benefit but at the same time present potential safety risk. Useful safety principles and practices have emerged from different quarters of the maritime sector as well as adjacent domains and the value in all perspectives should be sought and exploited. AI and particularly machine learning have found niche applications aboard conventional vessels and, as autonomy is increasingly adopted, then more reliance will be placed on existing AI applications and further applications will emerge. For instance, AI provides field-leading edge to voyage and energy optimisation, data-driven condition-based maintenance and sensor processing for assisted navigational situational awareness. With its numerous and disparate applications beyond the maritime domain, the safety of AI has developed a broader interpretation than prevention of physical harm. However, the development of principles and practice outside of the domain have useful read-across to autonomy at sea that should be adopted.

6. REFERENCES

1. The European Maritime Safety Agency, 'Annual Overview of Marine Casualties and Incidents 2023', 2023.

2. AEGIS consortium, 'Methodology for Safety and Security Analysis', *Deliverable D5.5 - Version Final, Horizon 2020 research and innovation grant 8599922023,* 2023.

3. RAJABALLY, E., and WYLIE, M., 'Maritime Autonomy and Safety at Sea', *Proceedings of the IMarEST's International Naval Engineering Conference and Exhibition*, 2024.

4. US National Transportation Safety Board, 'Collision Between Vehicle Controlled by Developmental Automated Driving System and Pedestrian, Tempe, Arizona March 18, 2018', *Accident Report NTSB/HAR-19/03 PB2019-101402*, 2019.

5. DNV (Det Norske Veritas), 'Testing of RBAT on specific cases of mass concepts, Report 4', *Report No. 2022-0481, Rev. 0*, 2022.

6. BOLBOT, V., and THEOTOKATOS, G., 'Risk assessments, fail-safe procedures and acceptance criteria', *Deliverable D2.4 of AUTOSHIP WP2*, 2021.

7. Lloyd's Register, 'LR Code for Unmanned Marine Systems', *ShipRight Design and Construction Additional Design Procedures*, 2017.

8. DNV (Det Norske Veritas), 'Autonomous and remotely operated ships', *Class Guideline DNVGL-CG-0264*, 2021.

9. Bureau Veritas, 'Guidelines for Autonomous Shipping', *Guidance Note NI 641 DT R01 E*, 2019.

10. American Bureau of Shipping, 'Guide for Autonomous and Remote Control Functions', July 2021.

11. MARITIME UK, 'Maritime Autonomous Ship Systems (MASS) UK Industry Conduct Principles and Code of Practice', *Version 7*, 2023.

12. HAWKINS, R., OSBORNE, M., PARSONS, M., NICHOLSON, M., MCDERMID, J., and HABLI, I., 'Guidance on the Safety Assurance of Autonomous Systems in Complex Environments (SACE)', *Assuring Autonomy International Programme (AAIP), University of York, Version 1*, 2022.

13. ISO (International Standards Organisation), 'ISO 21448:2022 Road vehicles - Safety of the intended functionality', *Edition 1*, 2022.

14. DAVIES, M., and BIRTWISTLE, M., 'Regulating AI in the UK, Strengthening the UK's proposals for the benefit of people and society', *Ada Lovelace Institute*, 2023.

15. UK Government, 'Introducing the AI Safety Institute', *CP 960*, 2023.

16. HABLI, I., and MCDERMID, J., 'AI Safety: Navigating the Expanding Landscape of Potential Harms', *Safety-Critical Systems Club Newsletter, 32*(2), 2024.

17. LESLIE, D., RINCÓN, C., BRIGGS, M., PERINI, A., JAYADEVA, S., BORDA, A., BENNETT, SJ., BURR, C., and FISCHER, C., 'AI Safety in Practice' *The Alan Turing Institute*, 2024.

18. UK DSTL (Defence Science and Technology Laboratory), 'Assurance of Artificial Intelligence and Autonomous Systems, A Dstl Biscuit Book', *DSTL/PUB136185*, 2021.

19. ISO/IEC (International Standards Organisation/International Electrotechnical Commission), 'Framework for Artificial Intelligence (AI) Systems Using Machine Learning (ML)', *ISO/IEC 23053 Edition 1*, 2022.

20. HAWKINS, R., PATERSON, C., PICARDI, C., JIA, Y., CALINESCU, R., and HABLI, I., 'Guidance on the Assurance of Machine Learning in Autonomous Systems (AMLAS)', *Assuring Autonomy International Programme (AAIP) University of York Version 1.1*, 2021.

21. UK DSIT (Department for Science, Innovation and Technology), 'Introduction to AI assurance', 2024.

22. UK IET (Institution of Engineering and Technology), 'The Application of Artificial Intelligence in Functional Safety', 2024.

23. US NIST (National Institute of Standards and Technology), 'Artificial Intelligence Risk Management Framework (AI RMF 1.0)', *NIST AI 100-1*, 2023.

24. UKRI, 'Report on the Core Principles and Opportunities for Responsible and Trustworthy AI', *Ref: PS22477*, 2023.

7. AUTHORS' BIOGRAPHIES

Eshan Rajabally is Maritime Autonomy Technology Lead at BMT. He is responsible for identifying and seeding technology based BMT consultancy and other service offerings in the field of maritime autonomy. The Author's previous experience includes road-mapping the Rolls-Royce vision for commercial uncrewed shipping and leading the subsequent development of several supporting technologies. This resulted in field-leading prototype demonstrators in autonomous path planning, remote engine room monitoring, navigational situation awareness and intelligent platform management.

Matt Wylie is Senior Safety Consultant at BMT. Here he provides safety assurance expertise to several clients, working with novel and emerging technologies. Matt has extensive experience developing safety cases for autonomous and remote-control surface ships, particularly in the risk assessment phase of safety case development.

Esin Turkbeyler is Senior Principal Artificial Intelligence Lead at BMT, with over 24 years' experience in the domain. During this time Dr Turkbeyler has developed several AI based solutions in Computer Vision for Defence and commercial markets. Within her current role, Dr Turkbeyler has most recently been working on Multi Agent Reinforcement Learning (MARL) algorithms, applied to autonomous cyber defence decision making.

LOW LATENCY VIDEO COMMUNICATION AT 5G NTN FOR SHIP REMOTE CONTROL

I Tsukagoshi, NIPPON KAIJI KYOKAI, Japan

SUMMARY

In recent years, Low Earth Orbit (LEO) satellites have started their commercial services, and there have been improvements in communication bandwidth and latency, compared to conventional ones using geostationary (GEO) satellites. Terrestrial mobile 5G significantly improves network performance compared with conventional LTE, which is extending its coverage to the sea, backcountry, and space, called NTN.

To remotely control MASS from ROC, it is essential to achieve adequate data delivery performance, along with the network capability ensured according to the connectivity underway. We consider the potential requirements for video data transmission so that the remote operator can recognize the ship's status by visual means. G2G latency throughout the video path is primarily caused by video codec and transport.

In this paper, we propose a transmission architecture that takes into account high efficiency, low latency, and reliability, also conduct simple latency simulation with NTN, and discuss issues for future transmission schemes.

Keywords: Remote control of MASS, Connectivity, LEO, 5G, NTN, Latency, Efficiency, Video Codec, G2G latency, VVC, GDR, ARQ, FEC, Reliable transport, Channel bonding

1. INTRODUCTION

Maritime Autonomous Surface Ship (MASS) development is underway in the maritime industry. In Maritime Safety Committee (MSC) of the International Maritime Organization (IMO), the Degrees of Autonomy (DoA) is organized as preliminary definition, where remotely controlled ship operated from another location is included [1]. Remote execution of ship is being considered as one of Concept of Operation [2]. In MEGURI 2040 Consortia of Japan, remote control has been applied in its experimental demonstrations [3].

The communication system of the MASS is divided into two categories: ship onboard communication which collects information from onboard equipment including sensors, radars, and so on, and ship-to-shore communication which transmits and receives data between the ship and the shore. This paper focuses on the latter.

When the ship is operated remotely, it is practical that it is connected to a Remote Operation Center (ROC). Remote operation from the ROC is considered necessary not only for remote ship operations during normal operations, but also for emergency overrides during the autonomous navigation of the ship.

The communication infrastructure envisioned for ship remote operations is based on terrestrial mobile 5G, a communication standard that significantly improves performance with respect to higher speed, and lower latency compared to conventional LTE. And 5G is extended to cover the sea, backcountry, and space, called NTN which is considered to enable efficient data transmission between MASS and ROCs.

For remote control of MASS, it is essential to ensure network performance according to the connectivity underway and to achieve adequate performance in data transmission. Among various sensor data onboard, video and audio from onboard cameras should be provided so that the ROC operators can recognize the ship's status visually and audibly. Video data requires large bandwidth, and low latency transmission is required for ROC operators to quickly assess live conditions around the ship. Moreover, the feedback commands from the operator should be delivered as fast as possible so that they are reflected to the onboard equipment immediately. Thus, real-time performance of two-way communication is required. This paper decomposes Glass-to-Glass (G2G) latency into individual factors and discusses about the reasons and measures. An architecture that considers high efficiency encoding, low latency, and reliable transmission from MASS to ROC is proposed. Simplified latency simulations with GEO and LEO satellites will be performed, and issues for investigation of transmission schemes will be presented.

2. COMMUNICATION BETWEEN MASS AND ROC

2.1 NTN, AN EXTENSION OF 5G

Communications between MASS and ROCs are considered to be switched between terrestrial and satellite communication networks as appropriate depending on the navigation phase. In the river, port, and coastal areas, it is assumed that the terrestrial radio network will be used while switching as appropriate, and that only satellite communications will be used when navigating in the open ocean.

Figure 1 shows the overall concept of connectivity in NTN between MASS and ROC. The satellite is connected to the ship by service link, and to the communication gateway by feeder link. Terrestrial mobile 5G is connected to the communications gateway. The 5G network is defined in the domains of "Radio Access Network (RAN)" (#1), "Core Network (CN)", and interfaces with Data Networks (DN) such as the Internet or LAN. As a simple configuration, frequency conversion and power amplification are performed in a relay function installed inside the satellite. Other configuration could be that some part of RAN (#1) functions is installed in the satellite, and in this case, direct radio access by the onboard User Equipment would be performed. In this way, the RAN domain of 5G network is extended to reach non-terrestrial areas, called as NTN (Non Terrestrial Network) [4].

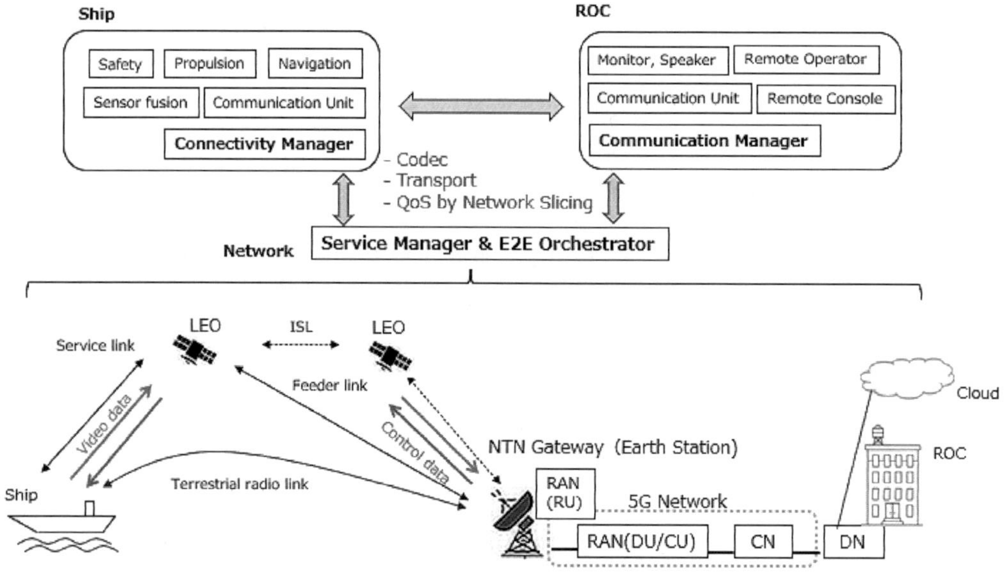

Figure 1: MASS-ROC connectivity in NTN

2.2 NETWORK PERFORMANCE

Network performance in 5G has targeted to greatly improve both throughput and latency compared with conventional LTE. Peak rate targets 20Gbps (Downlink) and 10Gbps (Uplink), which is more than 10 times higher than the conventional one. Latency is targeted to 1 millisecond or less in the radio section. As the latency LTE is said to be 10 milliseconds, it is expected to reduce to about 1/10 of the conventional one [5].

Although it is expected to improve performance, the End-to-End (E2E) network service has remained at the level of "aiming the target but not guaranteeing." In some applications such as Video on Demand (VoD) service, there is no problem. However, in certain time-sensitive applications such as tele-medicine and tele-operation, communication between the remote side and the local side is required super low latency in millisecond order accuracy. Conventional best-effort services without performance guarantees cannot cope with these.

The "network slicing technology" introduced in 5G Stand Alone, which is a full set function of 5G, logically divides the physical network by virtualization into slices, and the network performance can be customized for each of the "slice". The priority control of slices based on the pre-defined target could enable QoS (Quality of Service) control, so can guarantee qualified level of network service [6].

(#1) RAN consists of base stations that transmit and receive radio waves to and from User Equipment (UE) and perform handover control. The RU (Radio Unit) handles of radio physical layers, DU (Distributed Unit) has the functions of modulation, demodulation, FEC, and signal separation and synthesis, and CU (Centralized Unit) is responsible for radio resource control and packet data processing.

Though NTNs using satellites and other devices will add some constraints on 5G terrestrial performance, the expectation in NTN is that the targeted data is delivered in priority to other data over the E2E path, which could be enabled by the priority control done in 5G network slicing.

2.3 CONNECTIVITY

Communication connectivity is an essential component to achieve remote control of MASS. Sensor information and control data are required to be shared between MASS and ROC, which is supported by multiple systems in bi-directional, accurate, scalable, and redundant communication to minimize the risk of disruption [7]. To ensure the connectivity, the most important thing is the QoS required for each use case that is measured via fundamental KPIs such as latency, throughput, availability, and so on [8].

As shown in Figure 1, the Connectivity Manager on the MASS plays a fundamental role in the negotiation between the Service Manager of the network and the Communication Manager of the ROC to ensure the required service quality by exchanging Service Level Agreement (SLA) which is a contract between users and service providers [9]. The network's Service Manager checks with the E2E Orchestrator to see if it is feasible, and the E2E Orchestrator performs end-to-end data prioritization, route assignment, and network capacity allocation by managing RAN through CN. Along with the agreed performance, it is significant to develop appropriate data transmission formats in the upper layer of architecture.

3. VIDEO DATA TRANSMISSION

For the remote operation of MASS, larger bandwidth is desired to manually control the ship at ROC. It requires the transfer of relevant navigational data from the ship including high resolution video together with other sensor data. One of the technical challenges is the quality of available communication systems and how these systems can be used more efficiently to satisfy the requirements [10]. In terms of communication bandwidth, video data is considered dominant. It is necessary to consider highly efficient video encoding methods to enable ROC operator to judge the situation by visual hearing.

3.1 EVOLUTION OF VIDEO COMPRESSION TECHNOLOGY

Video compression is a technology that achieves to compress data by removing redundancy from video data, and consists of a combination of two major concepts:

- Spatial correlation: As a characteristic of image quality, adjacent pixels are very similar in natural images. Picture is transformed to the frequency domain, where the picture is analyzed and the low-frequency portions are intensively transmitted, while the high-frequency portions are rounded to zero. The rationale for this process is that human visual sensitivity is relatively low for high frequencies, and by rounding to zero, the amount of information can be reduced.
- Temporal correlation: The correlation between pictures is noticed and the redundancy of adjacent pictures is utilized. Redundant components are removed by inter-frame prediction with motion vectors.

De-correlated data by the above processes is subjected to the entropy coding which assigns a small amount of information to a high occurrence probability based on information theory [11].

Video CODECs (COmpression and DECompression) have evolved with the advancement of technology and have improved coding efficiency by incorporating various new compression tools with each generation.

The state-of-the-art codec internationally standardized at this time is VVC (Versatile Video Coding). How they have evolved is evident in the comparison of coding efficiencies between codecs to obtain equivalent image quality:

AVC|H.264 (170%) ⇒ HEVC|H.265 (100%) ⇒ VVC|H.266 (60%) when HEVC as the comparison criteria [12]. The smaller the value, the higher the compression performance. Note that the encoding performance depends on the texture of input video, and the universal performance is not guaranteed. The core message is the evolution of the encoding efficiency enables video transmission with higher picture quality within the bandwidth. In other words, the transmission bandwidth can be utilized more efficiently to achieve equivalent reconstructed picture quality.

In MASS-ROC transmission, a higher compression ratio means more options could be taken for remote operation in terms of usage of bandwidth, video quality, number of camera views and so on.

3.2 GLASS TO GLASS LATENCY

"Glass to Glass (G2G) latency" in video path refers to the time it takes for the video input from the image sensor on the camera to the final rendering on monitor, through video encoding, sent from the encoding system to the network, transmission in the network, received by the decoding system, video decoding, and displayed on the monitor, as shown in Figure 2.

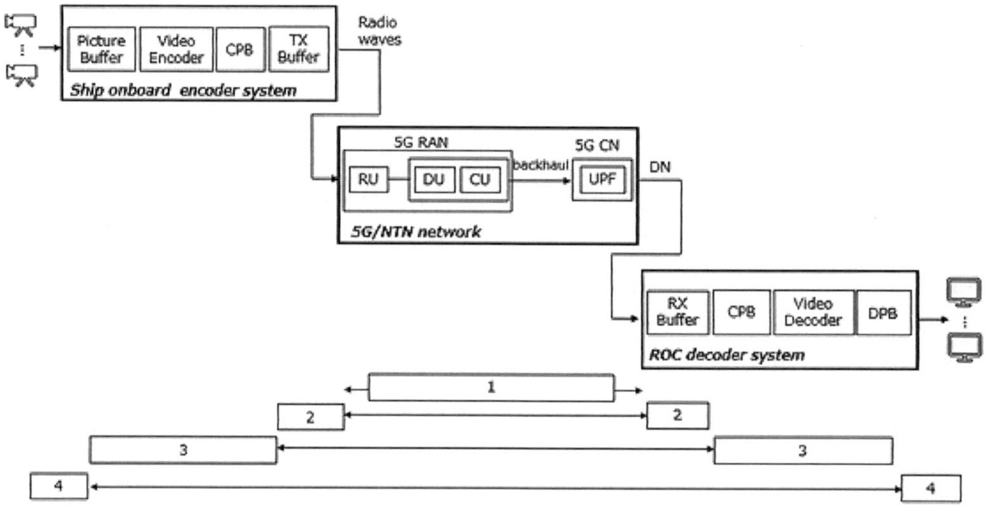

Figure 2: G2G latency

G2G latency is broken down into layers as follows:

1. Physical and Data Link Layer
   Latency of network processing, routing in RAN, CN, and propagation in distance.
2. Transport Layer
   Latency caused by retransmission buffer and/or FEC processing at TX Buffer and RX Buffer.
3. Codec layer
   Latency of picture reordering and smoothing buffer at Picture Buffer, CPB, and DPB.
4. Capture latency, and display processing latency
   Latency of in-camera capturing, image processing, and media player handling to display.

3.3 PICTURE REORDERING

When encoding video, video input sequence is divided into three types of pictures, Intra-picture which does not make inter-frame prediction, P-picture which refers to past picture, and B-picture which refers to past and future pictures. As shown in Figure 3 (1), the advantage of B-picture bidirectional prediction is that it can predict the part of the occlusions that appear and disappear in the picture, thus improving compression efficiency. In addition, since the two reference blocks are averaged, there is a noise reduction effect on the reference pixel signal, which reduces the prediction residual power and contributes to improve compression efficiency. Bidirectional prediction is one of fundamental coding tools supported across generations of video codecs [13].

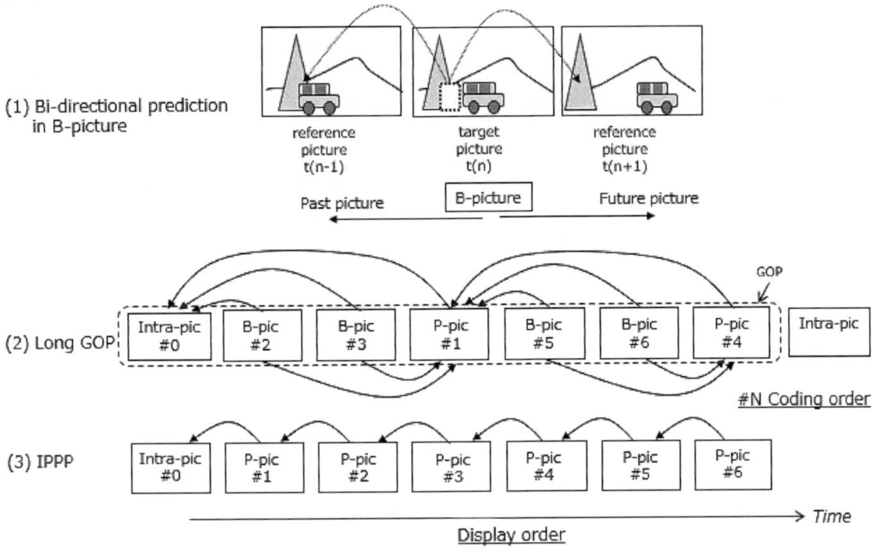

Figure 3: Bi-directional prediction and Picture Reordering

In B-picture encoding, a certain number of input signals are stored in frame memory "picture buffer" of Figure 2 to refer past and future pictures, and the input picture sequence is reordered, so-called "picture reordering".

Figure 3 (2) shows an example of a Group Of Pictures (GOP) containing Intra-, P-, and B-pictures. Because distance between the two Intra-pictures is long, it is called Long-GOP. The input picture sequence to encoder is in display order, and prediction is done by reordering the pictures in the "coding order" indicated by #N. The bitstream is organized in the coding order, and in the receiver, the decoder processes the pictures in that order. The decoded pictures are stored in DPB (Decoded Picture Buffer) to be predicted by subsequent pictures, and at the output of the buffer, pictures are reordered to display order. Both the encoder and the decoder need to store pictures in a certain period of time for picture reordering, which causes a latency in frame units. Compared to Long-GOP, Figure 3 (3) shows an example of an encoding sequence without B-pictures, as picture reordering is not needed, coding order and display order match.

3.4 SMOOTHING BITS

The size of encoded picture generated each picture greatly varies across pictures, because the complexity of the texture of picture varies on the captured scenes, and the encoding type varies across pictures.

While the size of Intra-picture (I-picture) is several times larger than the target bit rate because it encodes the texture of a picture, inter-picture (P-or B-picture) size is very small because it encodes the predicted residual components. If the encoded data is packetized and transmitted to network as it is, packets exceeding the target bitrate are either stuck within network or discarded at the relay node. To prevent this, the transmission data must be smoothed at the encoder's CPB (Coded Picture Buffer) output stage conforming to the target bit rate. Figure 4 illustrates this smoothing.

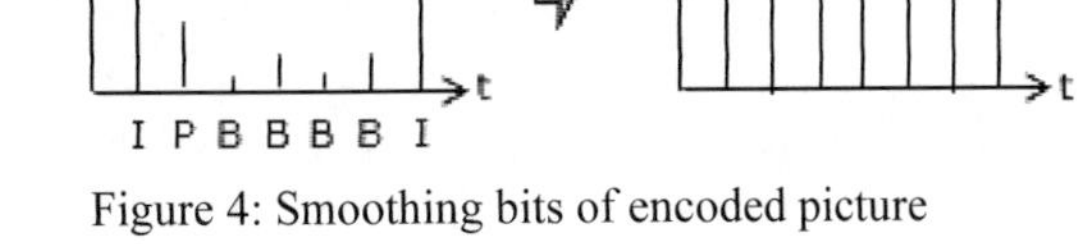

Figure 4: Smoothing bits of encoded picture

A concept of a "decoder reference buffer model" has been introduced. This is what the encoder takes into account the behavior of decoder's CPB by controlling the amount of generated bits at the encoder so that buffer failure such as Overflow/Underflow does not occur at the decoder's CPB.

As shown in Figure 5 (A), the vertical axis represents data amount and the horizontal axis represents time. The vertical space between Bit rate R(1) and Bit rate R(2) represents Buffer size A of CPB. The decoder receives the encoded bit stream at Bit Rate R and stores it in the CPB. The polyline represents the transition of buffering whose vertical part indicates the picture decoding. Its horizontal gridding represents the frame rate. After the first bit of the first encoded picture is input to the CPB, the first picture "Intra Pic" is decoded after the latency of Delay A. Decoding is always instantaneous, and the picture is deleted from the buffer. In the figure, buffer accumulation state is indicated by "Buffer occupancy." When the polyline exceeds Bit rate R (1), it becomes buffer underflow, and when it exceeds Bit rate R (2), it becomes buffer overflow. Overflow or underflow may lead to buffer collapsing, so the CPB in the encoder is managed by encoder rate control considering the CPB in decoder, so as not to happen. In practice, the Bit rate R of the target may adjust the encoding rate depending on the network status or the input video, and it is not always fixed but fluctuates in certain periods. Reading out from the CPB with the target bitrate in small intervals can be proceeded to packetization, so that it can be transmitted at the smoothed bit rate.

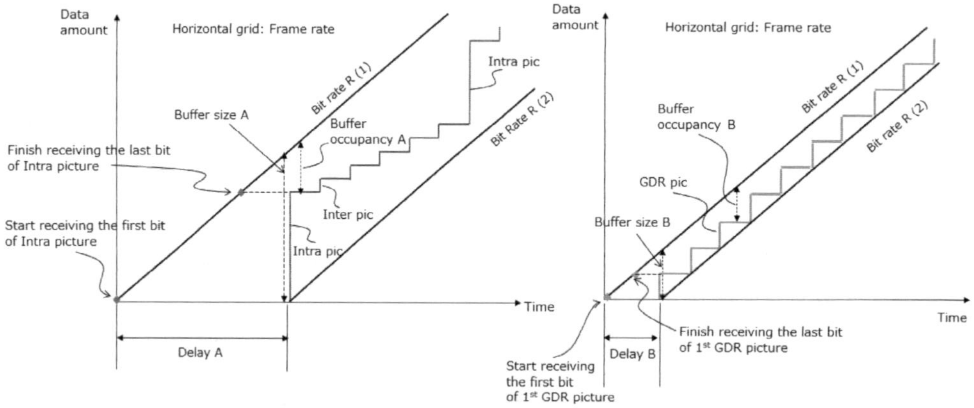

(A) Long-GOP (B) GDR

Figure 5: Decoder Reference Buffer Model

The decoder reference buffer model of Long-GOP type is shown in Figure 5 (A), and the latency for smoothing bits is indicated by Delay A which is not small. Note the Long-GOP has latency of picture reordering and smoothing bits, so its overall latency would be relatively high. To lower the latency, it would need to consider other coding scheme.

One of the methods to achieve this is the "intra column refresh", where the encoded sequence is composed of all inter-pictures. As shown in Figure 6, an "intra-slice" (cross hatched region), consisting of only intra-blocks, is placed every inter-picture and shifting its position in the consecutive pictures. The parts other than the intra-slice are encoded by inter-prediction referencing past pictures. Gray areas are correctly reproduced even at decoder startup, while white areas are correctly reproduced after one cycle of picture refresh. This feature is called GDR (Gradually Decoding Refresh), and signalling is defined as one of the essential functions in VVC codec [14].

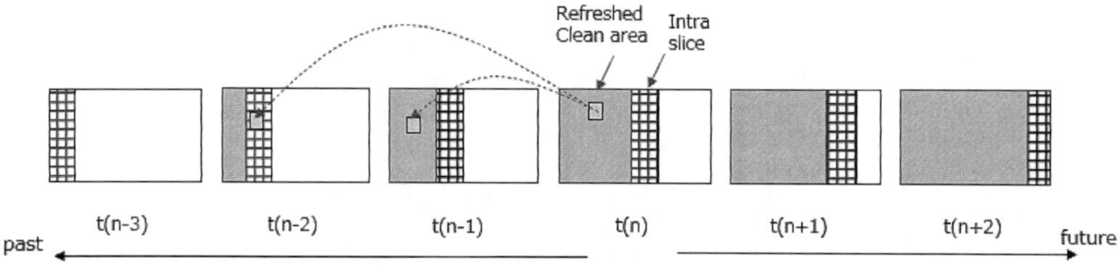

Figure 6: GDR by Intra column refresh

In GDR mode, since the area ratio of the intra slice portion to the non-intra slice portion is constant for each picture and all pictures are subject to the same encoding type, drastic variation in the encoded bits per picture is unlikely to occur. The decoder reference buffer for GDR is shown in Figure 5 (B). "Buffer size B" that can absorb small bits variation is somewhat necessary, but "Delay B" is obviously smaller than Delay A, since the difference in the generated bits across pictures is small. Thus, GDR can be used to avoid bitrate peaks caused by Intra-pictures, also as shown in Figure 6, the prediction is limited in GDR only to refer to the past picture, and since bidirectional prediction is not allowed, the latency of picture reordering does not occur. A small smoothing buffer with no picture reordering is the key to achieving low latency, which is an advantage for GDR.

On the other hand, the coding efficiency of GDR is not as high as that of Long-GOP because there are no bidirectional prediction and the compression tools are limited. To provide the equivalent picture quality as Long-GOP, the bit rate is expected to increase. The selection should be made based on the priority requirements of connectivity for MASS and ROC: picture quality with Long-GOP or low latency with GDR.

4. TRANSPORT FOR RELIABLE TRANSMISSION

Even if network performance is guaranteed in 5G and NTN, there is a possibility that packet loss occurs due to random errors caused by weather conditions or bursts of errors caused by congestion at the satellite transponders or routers along the way. When packet loss occurs during transmission, it affects to the decoding of compressed data, resulting in blocky errors in decoded picture or freezing of the picture, making continuous playback difficult. A mechanism to recover packet loss is required to get the reliable transmission, which is sometimes regarded as availability [15].

The processing for error recovery is performed in both TX and RX Buffers. There are two options to recover the lost packet: ARQ (Automatic Repeat reQuest) and FEC (Forward Error Correction).

ARQ : In networks with low packet loss, ARQ works effectively and has an advantage over FEC in terms of bandwidth gain and error recovery capability.

- No special calculation is required for both TX and RX sides.
- It is necessary to secure redundant bandwidth for retransmission.
- If the distance between transmission and reception is long and there is much packet loss, retransmission rate increases, which would cause traffic congestion.
- A latency corresponding to the size of the retransmission buffer occurs. Figure 7 (A) is an example with RX buffering of RTT*3 [16].

FEC: The sender side generates FEC packets by interleaving or block-aligned data, in either of rows or columns or both directions in the 2D array as shown in Figure 7 (B). The FEC packets are transmitted together with data packets. The receiver performs error detection, and when it detects packet loss, it will execute the error correction using FEC packets to recover the lost data. The figure describes that all lost data, P31, P32, P33, P34, P35, P45, P46, are recovered in order [17].

- There is no retransmission required, so it is suitable for real-time communication.
- Redundant bandwidth for FEC packet must be provided.
- The limit of correction capability is determined by the size of the FEC data.

In FEC, the appropriate error correction method differs whether the occurrence of packet loss is random or bursty, so it is expected to adopt an error correction method appropriate for the characteristics of transmission channel. Either ARQ or FEC, or combination of both, is selected depending on the overall requirements. If ARQ and FEC are combined, it is meaningful to design to what extent retransmission by ARQ is used and what type of data loss is corrected by FEC, depending on the bit error rate (BER) of the target transmission channel, its recovery performance, and latency requirements.

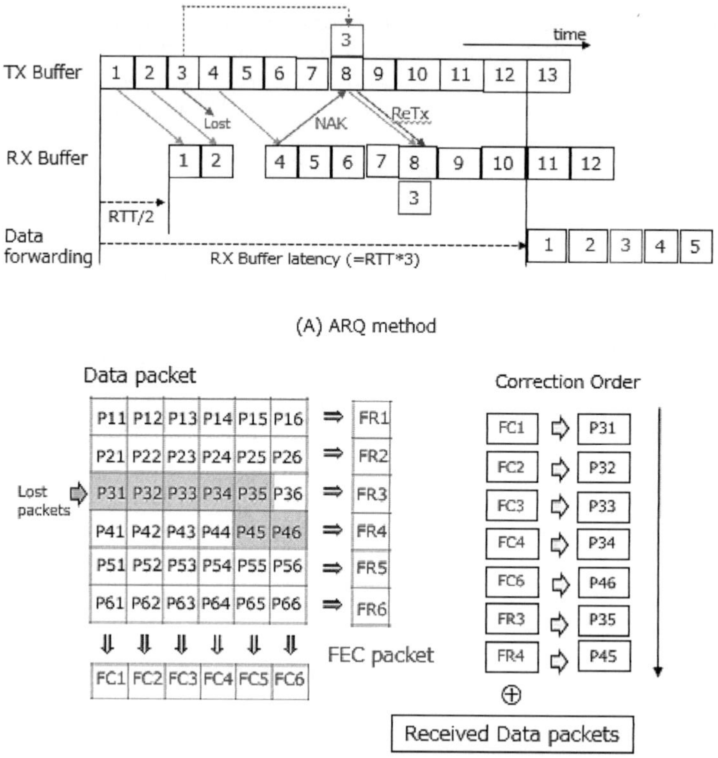

(B) FEC method

Figure 7: Error Recovery methods

5. PROPOSED VIDEO/AUDIO TRANSMISSION SYSTEM ARCHITECTURE

A protocol stack for a transmission architecture that satisfies the requirements of each layer described above is summarized and proposed as shown in Figure 8.

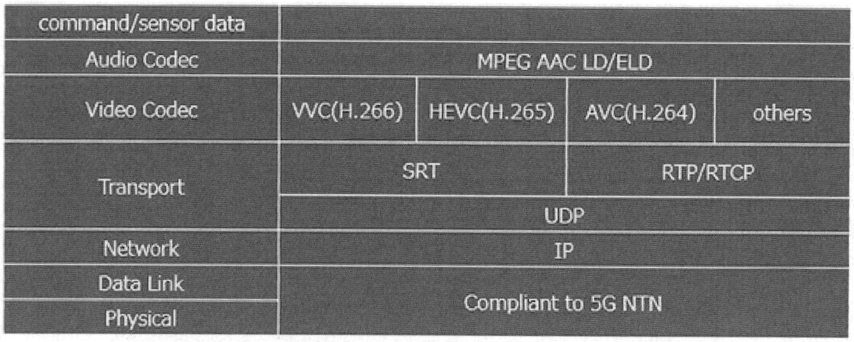

Figure 8: Proposed Transmission Architecture

① Physical layer and Data Link layer conform to 5G and NTN communication specifications.
② IP is appropriate for Network layer.
③ Transport layer is based on real-time transport using UDP, and on top of UDP, SRT(Secure Reliable Transport) or RTP/RTCP (Real-time Transport Protocol/RTP Control Protocol) would be appropriate as a container transport to send and manage synchronous data. Both protocols support ARQ and FEC.
④ Video codec is assumed to be either VVC, HEVC, AVC, standardized in ISO/IEC and ITU-T.

The highest compression efficiency is expected with VVC. Selection of appropriate encoding mode should be made depending on whether high efficiency or low latency is a priority. Audio codec might be appropriate to achieve low latency with MPEG AAC LD or MPEG AAC ELD.

6. LATENCY SIMULATION AND DISCUSSION

6.1 SIMULATION

For the upper layer of transport protocol, SRT or RTP/RTCP is recommended for real time transmission and error recovery to compensate packet loss. Error recovery scheme is ARQ or FEC supported by these protocols [18] [19].

In ARQ, when an error is detected in receiver, it automatically sends a NAK signal to request retransmission of the lost packet. The transmitter picks up the NAK target from its TX buffer and retransmits it. The time required for this bidirectional communication is measured in RTT (Round Trip Time), and the retransmission would be repeated N times till it gets to the receiver, which is the latency of ARQ. The value of N is determined by the recovery performance and the latency tolerance.

In FEC, the error correction is performed in the receiver. Though the latency of block interleaving and the arithmetic processing are added, overall latency is reduced compared with ARQ, because there's no retransmission. On the other hand, it is necessary to ensure a redundant space in band for sending FEC packets. FEC is considered valid within its error correction capability.

For experiments, it will be meaningful to assess the recovery performance of ARQ/FEC/ARQ+FEC depending on the transmission bandwidth including redundant bandwidth for error correction.

Prior to the experiment, as shown in Figure 9, G2G latency simulation is conducted on three encoding modes (Long-GOP/IPPP/GDR) and packet loss recovery methods with FEC and ARQ is plotted for LEO and GEO satellites. The vertical axis shows the latency in milliseconds, and the horizontal axis shows the relative bit rate comparison. The estimation was made under the following conditions.

- Frame rate of input video is 60Hz.
- Buffering at camera capturing and displaying for monitor is 1 frame period each.
- Long-GOP structure: In L=3 hierarchy, the Codec algorithmic latency is 10frames [20].
- Smoothing buffer size: 500 msec max (equivalent to satellite 4K broadcasting) [20].
- Retransmission buffer size: 4*RTT (for ARQ).
- RTT: 550msec (GEO), 50msec (LEO) (including network processing with vertical propagation latency).
- Long-GOP with ARQ is used as the bitrate standard.

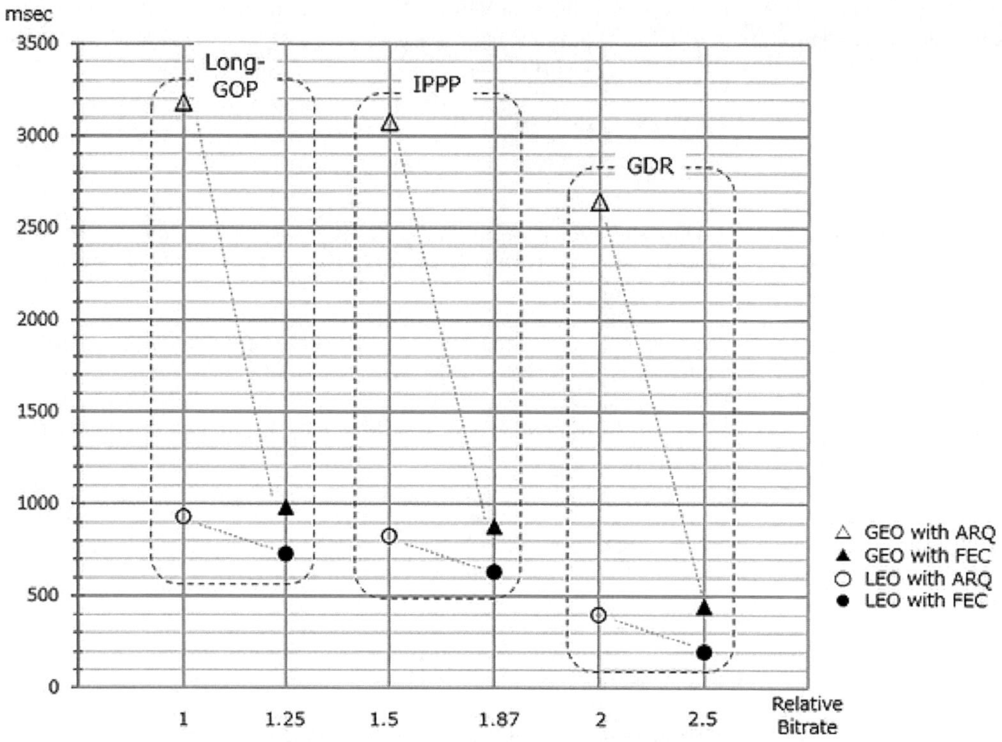

Figure 9: Estimated G2G latency

As can be seen in the figure, regarding the relationship between encoding efficiency and latency, Long-GOP is an efficient encoding mode using bidirectional prediction, but large latency occurs due to picture reordering and smoothing. "IPPP" differs from Long-GOP where picture reordering does not occur because bidirectional prediction is not performed. "GDR" is a low latency mode with no picture reordering and minimum smoothing latency. GDR has the lowest latency in the three encoding modes, however, the bit rate increases approximately twice as Long-GOP because of the limited compression tools available. IPPP is between Long-GOP and GDR in both image quality and latency.

Next, in the relationship between reliability and latency, the impact of ARQ is expected to be significant. The retransmission buffer is generally recommended to be about 3 to 4 times the RTT, but since it is a satellite link, it may be more practical setting the number N>4 times RTT. When encoded with Long-GOP mode, and ARQ is performed for LEO satellite, the plot of "LEO ARQ" is about 930msec, whereas when FEC is performed, the plot of "LEO FEC" is about 730msec. For comparison, when ARQ is performed for GEO satellite, the latency "GEO ARQ" is 3200msec, while FEC plotted as "GEO FEC" being approximately 1000msec. On the other hand, when encoded with GDR mode and ARQ is performed for LEO satellite link, the latency of "LEO ARQ" is 400msec. When FEC is performed, the plot of "LEO FEC" is 200msec. For comparison, when ARQ is performed for GEO satellite, "GEO ARQ" the latency is about 2600msec. When FEC is performed, the plot of "GEO FEC" is less than 500msec. For IPPP encoding mode, the latency is in-between Long-GOP and GDR.

6.2 DISCUSSION

From the simulation, when picture quality by high efficiency encoding is required and latency of 700~1000msec is allowed, Long-GOP can achieve high picture quality. When low latency is a priority, the latency could be reduced to 200~400msec with GDR, but it would cost to increase the bit rate.

When transmission reliability is included, in the case of LEO satellite use, the major difference in latency is more likely to be due to whether the coding mode is Long-GOP or GDR rather than ARQ or FEC. On the other hand, when using GEO satellite, the major difference in latency is largely due to ARQ retransmissions, and FEC would be preferred in a latency perspective. With respect to efficient use of bandwidth, FEC increases the bit rate with FEC packet transmission. In this simulation, 2D error correction is expected to increase the bit rate by about 25%, as shown in Figure 9. The efficiency of bandwidth usage will be reduced by that amount. If the error correction capability is to be increased, the number of FEC packets would increase and the ratio of redundant bandwidth to data bandwidth will increase. In extreme cases, redundant bandwidth and data bandwidth may be at the same level. If that is the case, there is another way for reliable transport by duplicating data packets, taking up a redundant channel space equal to the data bandwidth, without transmitting FEC packets. Figure 10 (A) shows a schematic representation of ARQ, FEC, and duplicated data transmission in the redundant band space. Verifying the relationship between transmission bandwidth and error recovery performance by these schemes would be one of the experimental targets. It is also key to study the error recovery schemes that combine ARQ and FEC with lower retransmission latency and to verify their usefulness.

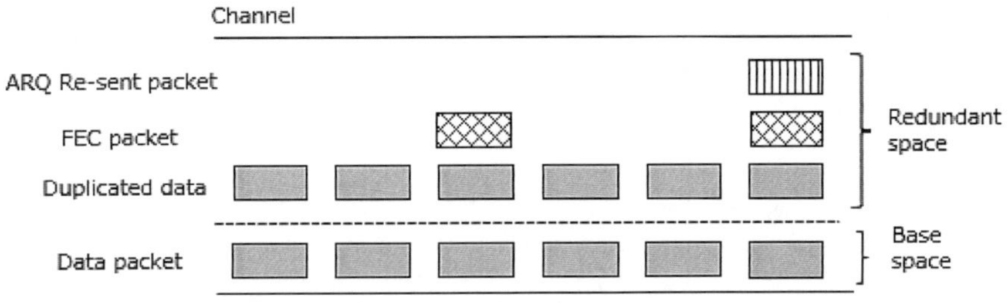

(A) Use of redundant channel space

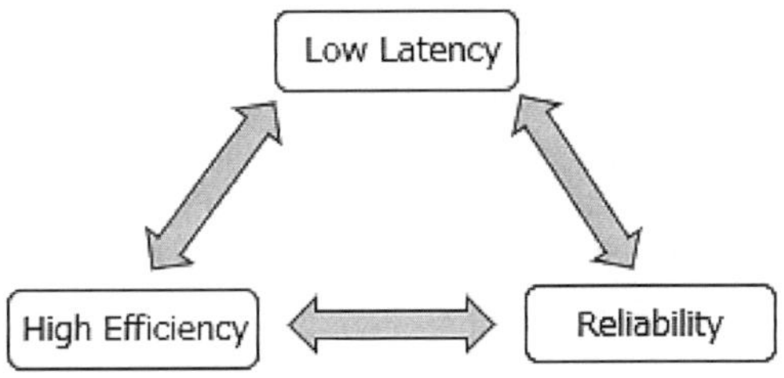

(B) Relationships of Performance

Figure 10: Use of redundant space and performance relation

It is expected to take measures according to the channel characteristics of NTNs. Appropriate encoding and transport schemes that satisfy the three requirements: High Efficiency, Low Latency, and transmission Reliability should be considered according to the channel characteristics. Given the priority to high efficiency of bandwidth utilization, it will impact on latency. Putting higher priority for low latency, it affects to the efficient use of bandwidth and reliability. For the reliable transmission, redundancy must be ensured which affects the effective use of bandwidth. The positioning of the three elements is triangular as shown in Figure 10 (B), and it will be necessary to decide how to balance them according to the situation.

This is where the codec compression performance comes into focus. The required bit rate of VVC, introduced at the beginning of this paper, is about 1/3 of that of AVC and 3/5 of that of HEVC, is expected to have higher compression performance, that is similarly applied to GDR. Higher efficiency encoding can reduce the required bit rate and increase free space. This can help alleviate channel congestion, so reduce packet loss and improve reliability. VVC also has a feature called RPR (Reference Picture Resampling) by which the resolution can be changed dynamically at any given timing [21]. For example, if you want to zoom in on a specific ROI (Region Of Interest), or if bandwidth gets narrowed due to the weather factors, you can seamlessly change the resolution HD <=> 4K <=> HD <=> SD, and so on.

These studies should be conducted to establish a transmission scheme for remote operation of MASS that is appropriate and balanced for the navigation phases.

7. OTHER CONSIDERSTIONS

7.1 EXTENDING UPLINK BANDWIDTH

For remote ship operation at ROC, real-time sharing of video and audio from multiple cameras onboard is considered essential. However, the uplink bandwidth from a ship to NTN satellite repeaters is not sufficient to send multiple video streams. By carrying out channel aggregation, channel bonding in other words, multiple channels can be bundled to increase bandwidth, which will contribute to allow more video streaming, or relaxing network load and thus reducing channel congestion, thereby enabling stable communications and reliable data delivery. It may also contribute to ensure seamless connectivity when switching between NTN and terrestrial networks such as 5G/LTE. The duplicated data transmission by channel bonding can be considered effective for redundancy. A similar trial would also be possible by applying to upcoming service with HAPS (High Altitude Platform Station).

7.2 STARLINK CHANNEL CHARACTERISTICS

Starlink provides commercial services using LEO satellites. Its link channel has been reported to experience significant fluctuations in performance on a regular basis [22]. The connection paths between UEs, satellites and earth stations are said to be reallocated at regular intervals. The measurements show that it causes latency and throughput fluctuations at the boundaries of 15 second intervals. If this phenomenon is common to LEO satellites, when conducting experiments on NTN, measures against periodic jitter and reduced throughput will need to be considered. It would be interesting to assess the performance of ARQ and FEC considering the time-varying characteristics of the target satellite link.

7.3 SAFETY NET FOR SHIP-LAND COMMUNICATION

Redundant MASS-ROC connections should be considered to ensure that communications between the two locations are not disrupted. As part of this effort, the possibility of securing communication with "satellite VDES" using LEO satellites as a safety net to maintain constant communications in the event of an anomaly in the redundant main line, although channel capacity is limited, is also worth considering for critical but minimum data sharing [23].

8. CONCLUSIONS

This paper described a method for low latency video communication from MASS to ROC. It is significant for ROC operators to have visual awareness of the ship's status to remotely control the ship. The network connectivity is assumed to be properly managed by QoS with network slicing in 5G, NTN as an infrastructure. On top of that we highlighted video codec and transport that dominate G2G latency and proposed a transmission architecture.

A simple latency simulation of the G2G latency over NTN satellites was conducted, and discussed a three-way relationship where high efficiency, low latency, and reliability are involved with each other. It will be important to take the right balance between these factors depending on the navigation phase of the ship and the situation during navigation. In the near future, we plan to experiment with actual satellite channels using a ship. At the same time, as an issue related to satellite service link capacity, uplink bandwidth from the ship is not sufficient at present, and it is necessary for both the communication infrastructure providers and users to make efforts for widening bandwidth. One of the approaches currently available for users is channel bonding. We believe this would be fundamentally meaningful to satisfy demands for redundant communications as well as for error recovery purpose to improve reliability. It may also contribute to ensure seamless connectivity when switching between NTN and 5G. In the experiment, we would like to verify the effectiveness of bandwidth extension by using ship underway and satellite link, as well as the discussed items.

Highly efficient and low latency transmission between ship and shore is a crucial objective, not only for remote operation of ships, but also for improving the well-being of the crew on board. The next generation of ship operations is expected to be safer, more efficient, and more attractive to seafarers.

We hope that this paper will serve as a reference for discussion on how to design transmission schemes for each communication use case between ship and land, including ship remote control.

9. REFERENCES

1. 'IMO takes first steps to address autonomous ships', May 2018, https://www.imo.org/en/MediaCentre/PressBriefings/Pages/08-MSC-99-mass-scoping.aspx

2. YAMADA, T 'Safety Evaluation for Technologies related to Autonomous Ships', *ClassNK Technical Journal No3*, 2021.

3. 'Results of demonstration tests of fully autonomous ship navigation on MEGURI 2040' MSC 106/INF.4 2 August 2022

4. SAAD, M M., et al., 'Non-Terrestrial Networks: An Overview of 3GPP Release 17 & 18', *IEEE Internet of Things Magazine,* January, 2024.

5. MOHYELDIN, E., 'Minimum Technical Performance Requirements for IMT-2020 radio interface(s)', *ITU-R Workshop on IMT-2020 terrestrial radio interfaces*

6. 'Commercializing 5G Network Slicing', A 5G Americas White Paper

7. HOLYHTYA, M., et al., 'Integrated 5G Satellite-Terrestrial Systems: Use Cases for Road Safety and Autonomous Ships'

8. LINDENBERGS, A., et al., 'Seamless 5G Multi-Hop Connectivity Architecture and Trials for Maritime Applications', *Sensors*, April 2023.

9. SERRANO, D., et al., 'Towards QoS-Oriented SLA Guarantees for Online Cloud Services', HAL open science, Feb. 2016.

10. RODSETH, Ø J., et al., 'Communication Architecture for an Unmanned Merchant Ship'

11. BAVAUD, F., et al., 'And Introduction to Information Theory and Applications' 2005.

12. SJOBERG, R., et al., 'Versatile Video Coding explained', *Ericsson Technology Review*, 2020.

13. CAGNAZZO, M., 'Video coding principles Introduction The hybrid coder MPEG standards', *Institut Mines-Telecom*

14. WANT, L., et al., 'Gradual Decoding Refresh for Versatile Video Coding', 2021 IEEE

15. ARIB TR-B34, 'Guidelines on Quality Control in Real-time Video Transmission over IP Networks for Program Production and News Gathering'

16. SHARABAYKO, M., 'SRT Protocol Overview'

17. BEGEN, A C., '1-D and 2-D Parity FEC', IETF73- November 2008.

18. 'SRT Packet Filtering & FEC' , Haivision/srt · GitHub

19. WATSON, M., et al., 'Forward Error Correction (FEC) Framework', *RFC6363*, October 2011.

20. 'Video Coding, Audio Coding, And Multiplexing Specifications for Digital Broadcasting ARIB Standard', *ARIB STD-B32 Ver. 3.11-E1*, July 2018.

21. BROSS, B., et al., 'Overview of the Versatile Video Coding (VVC) Standard and its Applications', *IEEE Transactions On Circuits And Systems For Video Technology, Vol. 31*, No. 10, October 2021.

22. MOHAN, N., et al., 'A Multifaceted Look at Starlink Performance', *Proceedings of the ACM Web Conference,* 2024.

23. 'Guideline: VHF Data Exchange System (VDES) Overview', IALA Edition2.0, December 2017.

10. AUTHORS BIOGRAPHY

Ikuo Tsukagoshi is a Senior Research Scientist at Research Institute of Nippon Kaiji Kyokai, Japan. He is engaged in communication technology at Ship-ROC for MASS Project from 2023.

In 1985, he received his Master's Degree of Electronics and Communication from the Faculty of Science and Engineering in Waseda University Graduate School, and joined Sony Corporation. He was a Visiting Scholar in Image Processing Systems Lab in the University of Washington, USA, at the behest of company in 1987-1988 where he conducted research on image compression techniques.

He had engaged in R&D on video processing and codec technology, development of media processor, DVD format, subtitle method, network game, and standardization of broadcasting technologies on UHDTV/3DTV for Japan/Europe/United States.

He is a member of IEEE, The Institute of Image Information and Television Engineers, and Information Processing Society of Japan.

AN INVESTIGATION INTO THE EFFECTS OF TRAINING ON THE FAULT DIAGNOSTIC SKILLS OF NAVIGATIONAL SEAFARERS

J P Chan, **R Norman**, **K Pazouki** and **D Golighty**, Newcastle University, United Kingdom

SUMMARY

The maritime industry is currently on the precipice for a technological revolution with the introduction of autonomous shipping. As developments with autonomous shipping progress this will impact for onboard operators and officers, requiring suitable training and competencies. As such, this paper introduces a study using a navigational bridge watchkeeping simulator and investigates the impact of training styles on the fault diagnostic skills of navigational seafarers, when presented with a fault in an automation system. For the study, 60 participants were shown one of two training videos on the technical or the behavioural skills of seafaring prior to the beginning of the bridge watchkeeping simulator exercise. The results were analysed using an Event Tree Analysis method to assess the performance of the participants and their diagnosis of a navigation fault. The findings of the study demonstrate that participants receiving the behavioural training video were more successful in diagnosing the fault in the exercise. Moreover, understanding the data gathered from the study could result in the development of future training regimes for navigational officers of the watch and offer a solution to optimise the evolution of the seafaring role.

1. INTRODUCTION

With the maritime industry looking towards the implementation of emerging technologies such as alternative fuels, digitalisation and autonomous systems, it is critical for all facets of the industry, in particular seafarers, to be versed with such developments to ensure their success [1]. Certain technological advancements will revolutionise daily onboard operations, specifically the relationship between autonomous navigation and the navigational officer [2]. Research has suggested that the development of autonomous technology will not eradicate the need for seafarers in the upcoming years [3, 4]. However, further research has indicated that there is a potential trust issue developing between the navigational officer and autonomous technology [5], which may result in significant hurdles in the introduction of autonomous technology.

The term autonomous shipping started to increase in popularity towards the end of the 2010's [6]. Initial discussions for autonomous shipping alluded to the possibility that such technology would ultimately be the demise of many seafaring roles and could offer ship owners various benefits resulting in a reduction in crew size [7]. However, research conducted since then has disproven the claims that autonomous shipping will spell the end for seafaring. Research initiatives analysing the human system interaction, such as human autonomous teaming (HAT) and onshore command centres (OCC), suggest that the direction of autonomous systems will result in a significant evolution for seafaring roles [8]. Moreover, throughout the course of the past 5 to 7 years, the International Maritime Organization (IMO) began their development of the maritime autonomous surface ships (MASS) working code, which has resulted in significant breakthroughs in terms of defining the levels on maritime autonomy and outlining the roadmap towards the implementation of autonomous shipping into the current global fleet [9].

With autonomous systems still being viewed as a future technology, it is critical that the maritime education and training (MET) sector looks to introduce autonomous technology training before commercialisation of the technology [10]. Research within the maritime industry has identified that maritime professionals do not believe that the modern International Convention on Standards of Training, Certification and Watchkeeping (STCW) framework, for training the officer of the watch (OOW), is suitable to incorporate MASS. Furthermore, as MASS progresses towards unmanned, remotely operated vessels, critical skills such as situational awareness (SA) and leadership require more focus in seafarer training [11]. Subsequently, as MASS are introduced, MET will have an obligation to support and ensure that the knowledge and competencies of the OOW training programme are not rendered obsolete by emerging autonomous technology [12]. Beyond the scope of seafaring, the vessel traffic services (VTS) will also be impacted by MASS with research identifying that individuals involved with the VTS believe that ship handling training thoroughly benefits the VTS operator in their daily operations and the inclusion of MASS training to MET will have a consequential effect on the training regime of VTS operators [13].

2. DEVELOPMENTS IN AUTONOMOUS SHIPPING TECHNOLOGY

2.1 MASS LEGISLATION

Work conducted by the IMO has highlighted the concerns of the industry, including the uncertainties behind machine learning and smart shipping [14]. As such the IMO have directed their future goals to address these concerns through further research prior to allowing MASS to operate within global waters and to develop a two stage legislation system with the primary legislation to regulate vessels adhering to MASS and the secondary legislation being a code of practice for remotely operated unmanned vessels [15]. Furthermore, the outlook of a fully autonomous maritime industry has garnered interest among the regulatory bodies that develop the guidance and rules for the industry. Members of the International Association of Classification Societies (IACS) have begun development on various aspects of MASS, in particular cyber-resilience [16, 17]. IACS have identified the lack of standards that are currently instated for MASS, and they are looking to participate in the development of MASS [18]. Furthermore, the individual members of IACS have begun to develop the foundations which will allow them to develop their own rule sets for MASS to adhere to [19, 20, 21]. Classification societies are also working together with various industrial bodies to conduct research into vessels and systems that will aid development of rules [22].

The development of autonomous systems and ships has created concerns within the maritime community regarding job security and has been a polarising topic among the industry [14, 23]. Research has shown that industry experts have many concerns regarding autonomous shipping including; that the reduction of crewing numbers, in addition to an increase in automation, will result in an over reliance on sensor technology; that the reduction of crewing numbers and the increase of automation will not result in a safe environment; and that the high levels of autonomy are not safe for human operation [7]. Such topics have subsequently complemented research regarding the manning requirements for vessels which has transitioned from minimum manning requirements to safe manning requirements, which denote the number of seafarers required to maintain both safety at sea and to determine the seaworthiness of the vessel [24]. Furthermore, it has been suggested that the removal of crew from an autonomous vessel may render the ship unseaworthy, due to the vessel being unable to comply with COLREGs which will have a further effect on the industry from a legislation perspective [25]. Vessels currently require a master onboard to act as the responsible human for the vessel and research into the removal of the master has highlighted that such actions require deeper investigation to determine the appropriate answer [26].

2.2 OPERATOR TRAINING & SKILLS

The standard route for British seafarers to become a fully qualified OOW and gain their unlimited, certificate of competency (CoC) has been laid out within merchant shipping notice (MSN) 1856 (M+F) Amendment 1 [27], and is summarised in Table 1. Research has shown that there are deficiencies with the training programme with individuals identifying a skill gap in the understanding of rules and misinterpreting the COLREGs [28].

The current maritime training regime identifies ten short courses as the minimum requirement for an individual to qualify as an OOW, as shown in Table 1. Beyond these requirements the STCW have developed many additional short courses to improve a seafarer's skills in maritime operations such as Bridge Resource Management, Maritime Crew Resource Management and Leadership and Teamwork [29]. However, many of the short courses offered are limited in spaces, diluted in information, optional and often costly for an individual to undertake [30]. Moreover, research has shown that that while short courses are beneficial for cadets to develop initial SA skills, many experts believe that further development of such skills are gained from onboard experience [31].

The current curriculum for the navigational OOW mandates that to obtain an OOW unlimited CoC, individuals must pass OOW written examinations from the Scottish Qualification Agency (SQA) at an approved SQA facility qualifying in both navigation and stability and operations [27]. Since the current OOW syllabus only covers two essential subjects, this presents an opportunity for MASS to be incorporated into the future curriculum. Yet with maritime institutions being limited by the approval of STCW legislation, institutions are unwilling to invest in training equipment for emerging technology [32]. Nevertheless, future training can include subjects beyond the technical aspects of onboard systems and the training and education of seafarers in subjects such as human factors, with an emphasis on SA, could benefit maritime professionals [33]. Moreover, research has shown that training in certain systems, such as the ECDIS, can have a positive effect on the SA of an individual, which may result in fewer maritime incidents occurring [34].

Table1. Requirements for OOW unlimited Certificate of Competency *[27]*

Requirements	Details
Minimum Age	18 Years Old
Minimum Seagoing Service	12 Months
Short Course Requirements	1. Personal Survival Techniques 2. Fire Prevention and Fire Fighting 3. Elementary First Aid 4. Personal Safety and Social Responsibility 5. Advanced Fire Fighting 6. Medical First Aid 7. Navigational Aids and Equipment Simulator Training (NAEST - Operational) 8. Efficient Deck Hand 9. Human Element, Leadership and Management (HELM - Operational) 10. Global Maritime Distress and Safety System (GMDSS)
Programme of Study	1. Honours Degree/Foundation Degree/Scottish Professional Diploma Route 2. Higher National Diploma (HND)/Advanced Diploma/Higher National Certificate (HNC)/Advanced Certificate Route
Medical Standards	ENG 1 Medical and Eyesight Test
Oral Examination	Pass

As the maritime industry looks to take the necessary steps towards MASS, research has identified that the MET sector can look to increasing the use of simulators within future OOW training programmes by delivering an authentic and practical education method to improve the training for individuals in aspects of the role of the OOW beyond vessel navigation [35, 36]. Moreover, further research has concluded that an immersive experience will accommodate the demands of the maritime industry and allow development of skills that are unable to be taught in a classroom setting [37]. The concept of utilising simulator training to replace sea going experience has already taken place in certain countries, with the Netherlands opting to allowing Dutch OOW students to obtain their CoC by completing 300 days of active seagoing experience and the remaining 60 days to be replaced by 15 days of bridge watchkeeping simulator training [38]. Despite simulators offering the OOW an environment to familiarise themselves with systems, research has shown that the seafarer's believe that the possibility of simulator work replacing first hand sea experience is not beneficial to the development of the OOW [31].Therefore, the criticality of this study is to assess the effects of different types of training to understand the impact that training has on the fault recognition skills of the officer of the watch.

The overall aim of the current study was to analyse the effects of different training packages and document their impact on the success rates in terms of both fault finding and fault diagnosis among navigational OOW. As such this paper assesses the performance of 60 participants, their overall reliance in onboard sophisticated automated systems and the impact of training topics has on the fault diagnostic skills of the OOW. To address this, aim the following hypothesis was identified, which was used to shape the direction and design of the study and develop a foundation exercise that could be used for future training courses in the maritime industry:

Hypothesis – Participants receiving a behavioural video training package, in comparison to a issuing participants with a technical based video training package including what systems will be impacted by autonomous shipping, will be more successful in recognising the fault with an automated navigation system.

3. METHODOLOGY

3.1 MATERIALS

For the study, it was critical that all participants had acquired a minimum of 12 months of navigational watchkeeping experience. This would then allow all participants to come into the bridge watchkeeping simulator and conduct a watch in accordance with Marine Guidance Notice (MGN) 315: Keeping a Safe Navigational Watch. It was also important to ensure that the bridge watch simulator set up closely simulated a working bridge watchkeeping environment, to increase the fidelity of the exercise. Therefore, various alarm sounds, including echo sounder alarms and doppler log alarms, were added to the simulation. Beyond the set-up of the bridge simulator, all participants would be issued a work pack containing the following items:

- Non-descriptive quizzes to imitate wheelhouse-based paperwork.
- A ship particulars worksheet;
- A vessel logbook.

The exercise was conducted using Kongsberg secondary bridge suites with the Kongsberg K-Sim® Navigation software. The vessel chosen for the exercise was the twin screw Maersk Majestic, 17,000 Twenty-foot Equivalent Unit (TEU) container vessel, which is defined as an Ultra Large Container Vessel (ULCV). The selection of this vessel meant that all participants had to be aware of their surroundings due to the vessel being larger than average. Participants were given sufficient opportunities to correct any fault which they encountered both aurally, through the clicking of the magnetic compass and alarms ringing, and visually, through cloud patterns moving and vessels deviating from traffic positions. Throughout the exercises, the participants had full communication access to the control station, which they could use if they began to encounter difficulties. Additionally, every participant was monitored using CCTV and microphones which were active throughout each exercise. For each participant, a watch handover was conducted during the mission brief. The handover consisted of telling the participants about the ship's position, traffic density and the weather conditions. During the exercise, the participant was the only person present in the simulator suite, unless they requested the presence of a look out. The parameters for both traffic vessels encountered in the exercise are shown in Table 2.

Table 2. Parameters of traffic vessels encountered in exercise

Vessel	Distance [Nm]	Bearing	Speed [knots]	Heading
1	7	45	20	110
2	18	300	12	240

3.2 TRAINING PACKAGES

The introduction of autonomous shipping and onboard systems has the potential to revolutionise the maritime industry, including the MET sector. Therefore, the intent of the design of the study was to determine whether the fault diagnostic skills of seafarers can be improved by altering the training received prior to beginning the exercise. As such, two different types of training information videos, both with a running time of 5 minutes, were delivered with the following information:

- Behavioural Training – A training package which included information regarding the behavioural skills that will need to be developed for autonomous navigation. The main topics delivered within this training package included:
  - Situational Awareness – What situational awareness is, how it can impact autonomy and how it can be improved.
  - Automation Bias – What automation bias is and how to combat it.
  - Complacency – The dangers associated with complacency and how not to fall victim to it.
- Technical Training – A training package which included information regarding the modern-day automated systems that may be used as the foundation for autonomous navigation. The main topics delivered within this training package include:
  - Autopilot – How the autopilot may be adapted for autonomous navigation.
  - ECDIS – How ECDIS can benefit future navigational autonomous systems.
  - Radar – The development of radar over time and how it can be adapted into a tool for autonomous shipping.

3.3 PROCEDURE

Upon entering the simulator suite, the participant would find a pilot card and ship particulars list on the workbench that would allow them to fill in their paperwork over the duration of the exercise. Once the exercise had begun, the vessel would be under autopilot control and the participant would then be allowed to alter the position of the vessel to suit their watch. Alarms were scheduled to sound every 5 minutes to replicate the sounds that the participant can expect on the bridge. However, after 12 minutes and 30 second the vessel would begin to deviate from its course due to a gyro drift failure. From the start of the failure the vessel would drift to port at 1 degree per minute until it achieved a 20-degree course deviation. If left unaltered the vessel would then sound an alarm at 25 minutes into the exercise to alert the participant that the cross-track limit had been exceeded, prompting the participant to investigate the fault. Should the participant identify or correct the fault then the time for the

participant was recorded as the number of seconds from the fault initiation to fault correction. Hence should a participant fail to identify the fault then that participant would receive a time score of 1050 seconds.
Figure 1(a) shows the radar plot at the beginning of the exercises, at a range of 12nm delivered by the gyro compass. As can be seen, the vessel's trajectory is set to travel along a course of 270 degrees and with one traffic vessel within its proximity. Furthermore, this is corroborated with the ECDIS view, at a range of 14nm, detailed in Figure 1(b). The ECIDS view details the speed of the vessel, the heading of the vessel regarding the gyro compass positioning and the GPS readings.

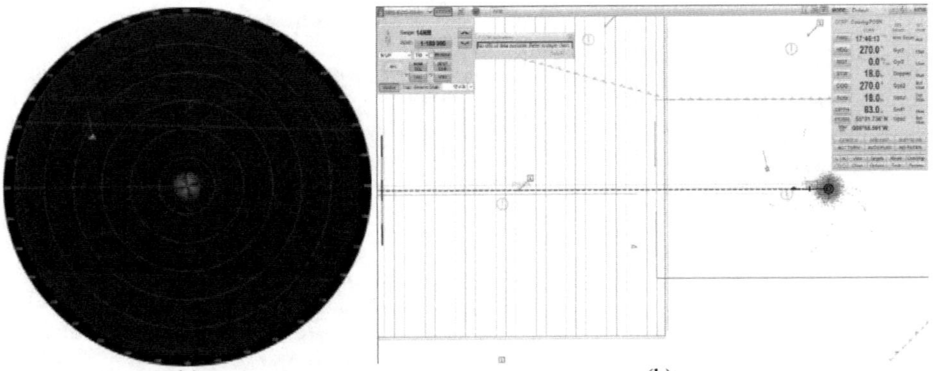

(a) (b)

Figure 1: Vessel Parameters from Radar and ECDIS View – Start

Figure 2 shows the displays for the end of the exercises if the vessel's navigation is left unattended, with Gyro 1 displaying a malfunction {Figure 2(a)}, and the correctly operating gyro compass, Gyro 2 {Figure 2(b)}. As can be seen from Gyro 2, if the participant does not correct the gyro drift fault, the vessel will have altered its course to a new heading of approximately 253 degrees. Gyro 1 shows the radar display that the participant could see if they did not alter the view to show Gyro 2, which would indicate to the participant that the vessel is still maintaining a 270-degree heading. The ECDIS plot shown in Figure 2(c) correlates to Gyro 2 from Figure 2(b). From the ECDIS view it can be seen that the vessel has deviated from its intended course to a new heading of 252.7 degrees. Additionally, no other parameters would be altered if left untouched by the participant.

(a) (b)

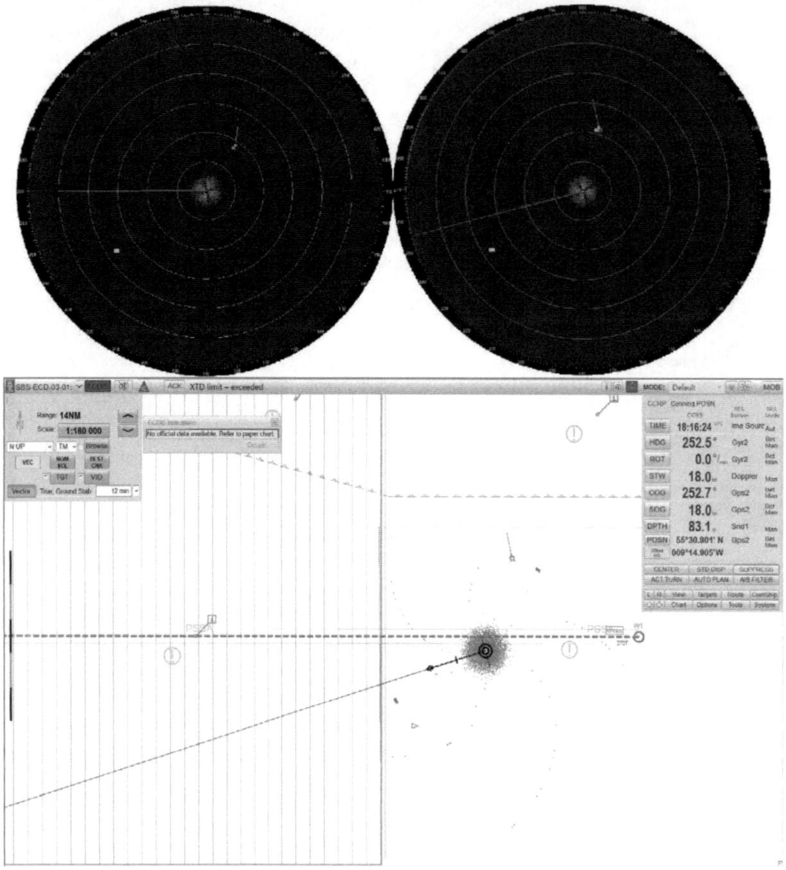

(c)

Figure 2: Vessel Parameters from Radar and ECDIS View – End

3.4 PARTICIPANT POOL

Sixty individuals participated in the study. Of these 60, 30 participants were fully qualified officers with 20 junior officers and 10 senior officers. The other 30 participants were unqualified officers, but with a minimum of 12 months sailing time as a navigational cadet. Most of the participants were set to be promoted within the months succeeding this study. Hence, the final phase navigational cadet participants would be qualified junior officers and junior officers would be promoted to chief mates. The groups were divided to define which received each training video. Participants were organised to ensure that there was an even split of 30 participants each that received the behavioural (*B*) and technical (*T*) training videos as shown in Table 3.

Table 3. Training Videos Received by Participant Ranks

Participant Rank	Qualified Seafarer	Training Video		Total Participants
		Behavioural	Technical	
Master	Yes	2	0	2
Chief Officer	Yes	3	5	8
2nd Officer	Yes	8	8	16
3rd Officer	Yes	2	2	4
Cadet	No	9	6	15
Rating to Officer	No	6	9	15
Total	-	**30**	**30**	**60**

4. RESULTS

Table 4 shows a summary of the participants success in both recognising and diagnosing the gyro drift failure at the conclusion of the exercise.

Table 4. Summary of findings

Training Received	**Fault Recognised**		**Fault Diagnosed**	
	Yes	**No**	**Yes**	**No**
Behavioural	15	15	2	28
Technical	3	27	1	29
Total	**18**	**42**	**3**	**57**

Due to using human participants, a multitude of event pathways were defined in the participants effort to complete the exercise. As such the findings of the study were analysed using an event tree analysis method, thus allowing for the identification of determining the individual's success in response to the event they encountered. Figure 3 shows the event tree for the exercise. The initial alarms at events 2 and 3 of the doppler log and echo sounder, allowed the participants to immerse themselves in a simulation replicating an active wheelhouse. Event 4 indicates the initiation of the fault. By event 5, nearly two thirds of the behaviourally trained participants had been proactive in addressing the course deviation, whereas only one third of the technically trained group had taken an initial course of action. Progressing through the exercise to event 9, this allowed participants to formally acknowledge the instance of a malfunction and document their findings of the exercise. From event 10, participants then can call the captain and alert them to the ongoing situation. As shown in the ETA, fifteen and three participants from the behaviourally and technically trained groups, respectively, successfully navigated the exercise to this event. All participants that had successfully addressed event 10 were successful in identifying the fault. However, both events 11 and 12 allowed the participant to further explore the fault with the aim to correct it. At event 11, five and three participants from the behaviourally and technically trained groups, respectively, successfully changed over the autopilot control from gyro compass 1 to gyro compass 2. However, it may be assumed that by the participants alerting the captain to the malfunction, it would ultimately be identified at a later point in a longer watch. Ultimately, from the cohort of 60 participants, two participants from the behaviourally trained group and one participant from the technically trained group were successful in their actions at event 12, resulting in a total of three participants following the correct procedures in diagnosing the fault.

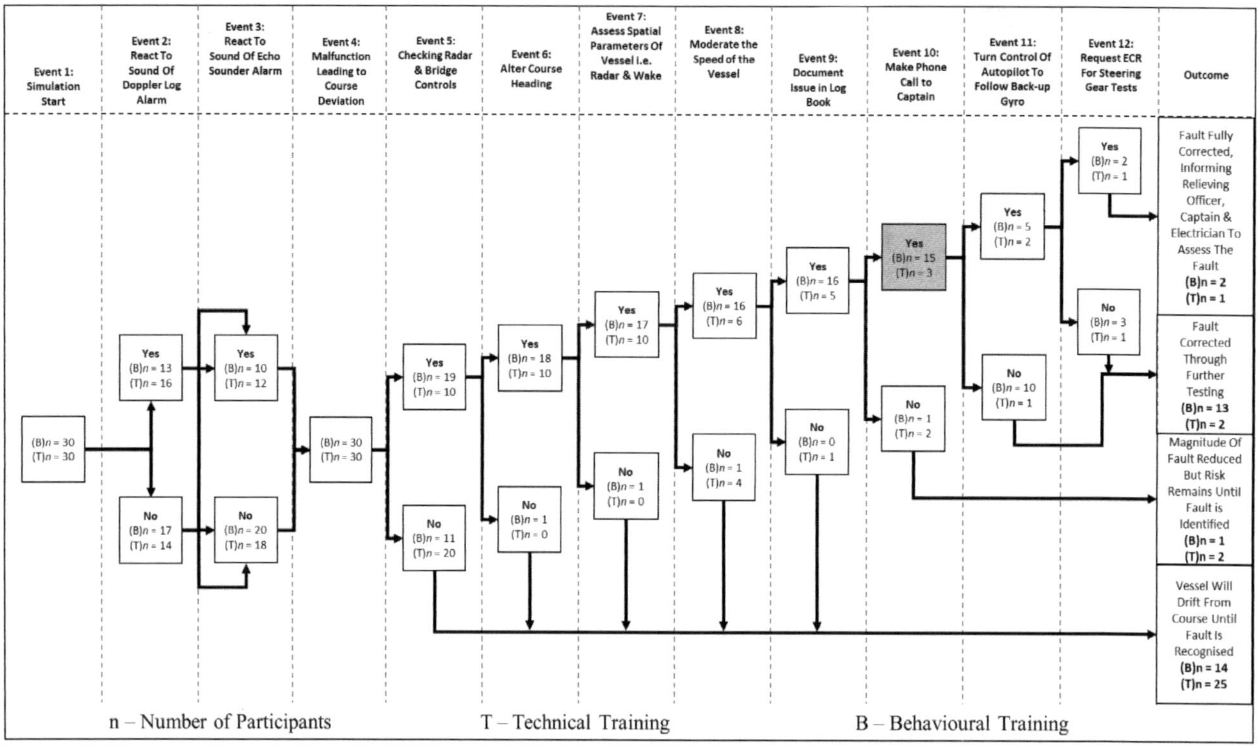

Figure 3: Fault Diagnosis Event Tree Analysis (ETA)

5. THE EFFECTS OF TRAINING PROGRAMMES

By implementing the ETA, it was possible to gain a greater understanding to the detail of how each participant performed during the exercises. The complexity of exercise allowed for a wide number of pathways to be taken by the participant. However, the use of the ETA regulated the safety barriers encountered by each individual and identified the performance of the participant at each event. Additionally, the ETA chart in Figure 3 shows that despite there being a variety of potential outcomes, an understanding can be developed of which event path was the most common among the participants. As such the findings of this study support the hypothesis that the participants receiving the behavioural training package outperformed the participants receiving the technical training package, in recognising the fault. Subsequently, when assessing the fault diagnostic skills of the participants it can be seen from Figure 3 that participants receiving the behavioural training package were more successful in diagnosing the fault. However, of all the participants, only three individuals successfully diagnosed the fault suggesting that, while behavioural training can benefit the OOW, further training would need to be undertaken to improve an OOW skill set to align with potential autonomous shipping requirements.

With Maritime 2050 on the horizon, the industry should aim to learn and develop from both ongoing autonomous maritime research and the introduction of autonomous technology in other transportation sectors. With the maritime industry looking to implement autonomous systems imminently, training the OOW will ultimately influence its success. As such, the development of this study aims to provide the maritime industry with the knowledge of how to adapt future navigational officer training to aid the evolution of the navigational seafaring role.

Research has already identified that the current curriculum contains outdated training aspects such as celestial navigation and paper chartwork and that cadets believe that the educational sector of the industry is not suitably arming them with the knowledge to succeed in the future of shipping [39]. The work conducted within this study has identified, with the transition towards autonomy, that behavioural training has the potential to be the most beneficial method for trainees. When comparing the success rates of finding the fault among participants, within their respective training groups, it was found that 50% of participants from the behaviourally trained group were successful in finding the fault, whereas 10% of technically trained participants could find the fault. The results of this study have shown that while the training of officers must be adapted for autonomy to succeed, future topics should focus less on the knowledge of the system and look to employ a more reflective method of education. While the behaviourally trained group were not infallible, the results of this study identify that incorporating aspects such as SA, AB and complacency into the maritime education curriculum can improve the fault recognition skills of participants. However, despite the participants receiving the training videos, a significant number of participants failed to correctly diagnose the fault they encountered.

With the direction of the maritime industry tending towards autonomous operations, similar studies have been conducted using small sample pools [40]. It was crucial for this study to develop a rich participant pool with a wealth of seagoing experience, thus the cohort assessed could be considered indicative of the current body of navigational seafarers. From the 60 participants assessed, 30% successfully recognised and reacted to the gyro drift failure that was causing a course deviation. However, only 10% of participants receiving technical training successfully reacted to the course deviation stemming from an automated fault, whereas 50% of the behaviourally training participants found the fault. Nevertheless, the findings of the study suggest that while additional training on key subjects is beneficial to the fault-finding skills, they have little benefit to the individual's diagnostic knowledge and skill set. Conversely, all participants have received sufficient training to undertake the responsibilities of a sole watchkeeper on the bridge, which raises the potential question of "does the average navigational officer experience an overwhelming level of cognitive workload to the detriment of the safety of the vessel?". Additionally, despite being given sufficient alarms and prompts throughout the exercise, a substantial number of individuals opted to ignore the warnings and potentially mistakenly place trust in the autopilot operating correctly.

6. CONCLUSION

With the introduction of autonomous shipping comes the conundrum of the industry attempting to identify the various topics that should be taught to seafaring students in future syllabi. By looking at past systems, for example ECDIS, the development of which was underway by the 1990s [41], there are still continual issues that result in maritime incidents due to a lack of understanding of the system [42, 43]. However, over the course of the next four decades many maritime accidents have been attributed to a lack of knowledge of the system and due to the multiple designs of the system there is a lack of familiarity [44, 45]. Tempering the expectations of autonomy and improving the understanding of technology for current officers offers the potential for a smoother transition to MASS. Beyond educating the OOW on future systems that currently have yet to be designed, the industry can focus on educating students on the behavioural aspects of seafaring, offering many officers the opportunity to hone skills such as fault recognition; fault diagnosis; and SA, in addition to learning about the hazards of AB; and complacency, resulting in a cohort of seafarers equipped with the problem solving skills to complement autonomous shipping.

The overall premise of this study was to ascertain the impact of prospective future training topics and their influence in fault diagnosis success. While the data suggests that behavioural training improves the fault diagnostic skills it appears that it is more suitable for fault recognition, as shown by the success of fault finding from Figure 3. As such, further developments can be made to optimise the behavioural training package to improve the number of participants successfully diagnosing the fault. Moreover, the data analysis of the study provided evidence that the success rate is dependent on which training package the individual received, as the behaviourally trained group outperformed the technically trained group. However, it is clear from the outcome of this study that more work needs to go into the development of training to equip seafarers with the skill set they need to futureproof their careers.

As the maritime industry transitions towards autonomous shipping this study has shown that training is of paramount importance and the key to its success. Nevertheless, through the developments of autonomous shipping it is evident that the maritime industry will require their training standards to undergo a complete overhaul, with the likes of paper chartwork and celestial navigation becoming potential subjects of the past. This study has shown that topics such as bias, complacency and situational awareness hold relevance in an industry that is steadily moving towards utilising more sophisticated systems to assist vessel operations.

7. REFERENCES

1. JO, S., D'AGOSTINI, E., and KANG, J., 'From Seafarers to E-farers: Maritime Cadets' Perceptions Towards Seafaring Jobs in the Industry 4.0', *Sustainability, vol. 12, no. 19*, 2020.

2. EMAD, G R., KHABIR, M., and SHAHBAKHSH, M., 'Shipping 4.0 and Training Seafarers for the Future Autonomous and Unmanned Ships', *Qeshm Island*, 2024.

3. SHAHBAKHSH, M., EMAD, G R., and CAHOON, S., 'Industrial revolutions and transition of the maritime industry: The case of Seafarer's role in autonomous shipping', *The Asian Journal of Shipping and Logistics, vol. 38*, no. 1, pp. 10-18, 2022.

4. MALLAM, S C., NAZIR, S., and SHARMA, A., 'The human element in future Maritime Operations – perceived impact of autonomous shipping', *Ergonomics, vol. 63, no. 3*, pp. 334-345, 2020.

5. AALBERG, A., HOLEN, S M., KONGSVIK, T., and WAHL, A M., 'Does it do the same as we would? How trust in automated shipboard systems relates to seafarers' professional identity', *Safety Science, vol. 172, no. 106426*, 2024.

6. ELLEFSEN, A L., ÆSØY, V., USHAKOV, S., and ZHANG, H., 'A Comprehensive Survey of Prognostics and Health Management Based on Deep Learning for Autonomous Ships', *IEEE Transactions on Reliability, vol. 68, no. 2*, pp. 720 - 740, 2019.

7. HANNAFORD, E., and HASSEL, E V., 'Risks and Benefits of Crew Reduction and/or Removal with Increased Automation on the Ship Operator: A Licensed Deck Officer's Perspective', *Applied Sciences, no. 11*, p. Paper 3569, 2021.

8. MANUEL, M E., and BAUMLER, R., 'The evolution of seafarer education and training in international law', *Maritime Law in Motion*, pp. 471-494, 2020.

9. International Maritime Organization, 'Development Of A Goal-Based Instrument For', IMO, 2024.

10. ABOUL-DAHAB, K M., 'The Readiness of the Maritime Education for the Autonomous Shipping Operations', 2021.

11. MALLAM, S C., NAZIR, S., and SHARMA, A., 'The human element in future Maritime Operations – perceived impact of autonomous shipping', *Ergonomics, vol. 63, no. 3*, pp. 334-345, 2020.

12. NARAYANAN, S C., EMAD, G R., and FEI, J., 'Theorizing seafarers' participation and learning in an evolving maritime workplace: an activity theory perspective', *WMU Journal of Maritime Affairs, vol. 2*, pp. 165-180, 2023.

13. JANSSEN, T., BALDAUF, M., and CLARESTA, G., 'From Ship to Shore – Studies Into Potential Practical Consequences of Autonomous Shipping on VTS Operation and Training', *TransNav - The International Journal on Marine Navigation and Safety of Sea Transportation , vol. 17, no. 2*, pp. 383-390, 2023.

14. KIM, M., JOUNG, T H., JEONG, B., and PARK, H S., 'Autonomous shipping and its impact on regulations, technologies, and industries', *Journal of International Maritime Safety, Environmental Affairs, and Shipping , vol. 4, no. 2*, pp. 17-25, 2020.

15. IMO - Maritime Safety Committee, 'Development of a Goal-Based Instrument for Maritime Autonomous Surface Ships (MASS)', IMO, London, 2022.

16. IACS, 'Cyber resilience of on-board systems and equipment (E27)', *IACS*, 2022.

17. IACS, 'Cyber resilience of ships (E26)', *IACS*, 2022.

18. IACS, 'International association of classification societies: Position paper – MASS', March 2019. [Online]. Available: chrome-extension://efaidnbmnnnibpcajpcglclefindmkaj/ https://iacs.org.uk/media/8673/iacs-mass-position-paper-rev2.pdf. [Accessed 18 May 2023].

19. American Bureau of Shipping, 'Autonomous Vessels', *Houston*, 2022.

20. Det Norske Veritas - Germanischer Lloyd, 'Remote-controlled and autonomous ships in the maritime industry', *Oslo*, 2018.

21. Bureau Veritas, 'Guidelines for Autonomous Shipping', October 2019. [Online]. Available: chrome-extension://efaidnbmnnnibpcajpcglclefindmkaj/https://erules.veristar.com/dy/data/bv/pdf/641-NI_2019-10.pdf. [Accessed 16 June 2023].

22. SCHIARETTI, M., CHEN, L., and NEGENBORN, R R., 'Survey on Autonomous Surface Vessels: Part I - A New Detailed Definition of Autonomy Levels', *Computational Logistics*, pp. 219-233, 2017.

23. RØDSETH, Ø J., and VAGIA, M., 'A taxonomy for autonomy in industrial autonomous mobile robots including autonomous merchant ships', Ulsan, 2020.

24. MACDONALD, R., 'Safe manning of ships — Yesterday, today and tomorrow', *WMU Journal of Maritime Affairs, vol. 5, no. 2*, pp. 143-151, 2006.

25. CAREY, L., 'All Hands Off Deck? The Legal Barriers to Autonomous Ships', in *NUS Law Working Paper*, 2017.

26. STĘPIEN, B., 'Can a ship be its own captain? Safe manning of autonomous and uncrewed vessels', *Marine Policy, vol. 148*, p. 105451, 2023.

27. Maritime and Coastguard Agency, 'MSN 1856 (M+F) - Amendment 1 - UK Requirements for Master and Deck Officers', 19 December 2022. [Online]. Available: https://www.gov.uk/government/publications/msn-1856-mf-amendment-1-uk-requirements-for-master-and-deck-officers. [Accessed 16 May 2023].

28. MOHOVIC, D., MOHOVIC, R., and BARIC, M., 'Deficiencies in Learning COLREGs and New Teaching Methodology for Nautical Engineering Students and Seafarers in Lifelong Learning Programs', *The Journal of Navigation, vol. 69*, no. 4, pp. 765-776, 2016.

29. STCW, 'STCW Courses', 2024. [Online]. Available: https://www.stcwdirect.com/stcw-courses/. [Accessed 28 February 2024].

30. HAUGHTON, C., 'STCW leadership development and management training in practice. Is leadership training a waste of money?', Manilla, 2011.

31. EVIDENTE, L G., KANNAN, K., LOGINOVSKY, V A., MORI, Y., and SHOSHIN, A R., 'The effect of preparatory bridge simulator training on development of situation awareness skills of watchkeepers', *Russian Journal of Water Transport, vol. 73, no. 4*, pp. 226-242, 2022.

32. EMAD, G R., and GHOSH, S., 'Identifying essential skills and competencies towards building a training framework for future operators of autonomous ships: a qualitative study', *WMU Journal of Maritime Affairs*, 2023.

33. BALYAN, R., and DHANKHER, D., 'Analyzing the Effective of Training Needs Analysis in Shipping Industry', *International Journal Of Innovations In Engineering Research And Technology, vol. 10, no. 3*, pp. 85-95, 2023.

34. BARIC, M., GRBIĆ, L., PERICIN, L., and JELIC, R., 'The Role of the ECDIS on the Development of Situational Awareness – a Study on Grounding Accidents', *TransNav - The International Journal on Marine Navigation and Safety of Sea Transportation , vol. 17, no. 1*, pp. 219-225, 2023.

35. TUSHER, H M., NAZIR, S., GHOSH, S., and RUSLI, R., 'Seeking the Best Practices of Assessment in Maritime Simulator Training', *TransNav the International Journal on Marine Navigation and Safety of Sea Transportation, vol. 17, no. 1*, pp. 105-114, 2023.

36. VIDAN, P., BUKLJAŠ, M., PAVIĆ, I., and VUKŠA, S., 'Autonomous Systems & Ships -Training and Education on Maritime Faculties', *Budva*, 2019.

37. DEWAN, M H., GODINA, R., REZAUL, M., CHOWDHURY, K., and MAN, M., 'Immersive and Non-Immersive Simulators for the Education and Training in Maritime Domain-A Review', *Journal of Marine Science and Engineering, vol. 11, no. 147*, 2023.

38. UITTERHOEVE, W., and LEUNEN, G V., 'Lifelong Learning: The Dutch Perspective on the Role of Simulators in Maritime Education and Training', *Scientific Journal of Gdynia Maritime University, vol. 119*, no. 21, pp. 54-63, 2021.

39. BOGUSŁAWSKI, K., GIL, M., NASUR, J., and WRÓBEL, K., 'Implications of autonomous shipping for maritime education and training: the cadet's perspective', *Maritime Economics & Logistics, vol. 24, no. Special Issue - Autonomous Shipping*, pp. 327-343, 2022.

40. PAZOUKI, K., FORBES, N., NORMAN, R A., and WOODWARD, M D., 'Investigation on the impact of human-automation interaction in maritime operations', *Ocean Engineering, vol. 153, no. 1*, pp. 297-304, 2018.

41. GREER, R A., 'The United States Navy Electronic Chart Display and Information (ECDIS) development project', Las Vegas, 1994.

42. Marine Accident Investigation Branch - MAIB, 'Report on the investigation of the grounding of Ovit in the Dover Strait on 18 September 2013', *MAIB, London*, 2013.

43. Marine Safety Investigation Unit, 'Safety Investigation Report MV UNIVERSAL DURBAN Grounding in position 02° 14.65' N 109° 05.17' E 13 May 2017', *Transport Malta*, 2018.

44. MAIB DMAIB, 'Application and usability of ECDIS', *MAIB & DMAIB*, 2021.

45. ŽUŠKIN, S., BRČIĆ, D., URODA, M., and STRABIĆ, M., 'Evolving ECDIS: Concept Development Through Different Manufacturer Models Comparison', *TransNav: The International Journal on Marine Navigation and Safety of Sea Transportation, vol. 17, no. 1*, p. 227, 2023.

46. LEE, J D., and SEE, K A., 'Trust in Automation: Designing for Appropriate Reliance', *Human Factors: The Journal of the Human Factors and Ergonomics Society, vol. 46, no. 1*, pp. 50-80, 2004.

47. FAN, L., CHING-HUNG, L., GANGYAN, X., CHUN-HSIEN, C., and LI, P K., 'A QFD-Enabled Conceptualization for Reducing Alarm Fatigue in Vessel Traffic Service Centre', *Advances in Transdisciplinary Engineering, vol. Transdisciplinary Engineering: A Paradigm Shift, no. 5*, pp. 821-828, 2017.

48. TAM, K., HOPCRAFT, R., CRICHTON, T., and JONES, K., 'The potential mental health effects of remote control in an autonomous maritime world', *Journal of International Maritime Safety, Environmental Affairs, and Shipping, vol. 5, no. 2*, pp. 40-55, 2021.

49. MASALONIS, A J., DULEY, J A., and PARASURAMAN, R., 'Effects of Manual and Autopilot Control on Mental Workload and Vigilance During Simulated General Aviation Flight', *Transportation Human Factors, vol. 1, no. 2*, pp. 187-200, 1999.

50. GHOLAMREZA, E., and WOLFF, M R., 'Contradictions in the practices of training for and assessment of competency: A case study from the maritime domain,', *Education + Training, vol. 50, no. 3*, pp. 260-272, 2008.

8. AUTHORS BIOGRAPHY

Dr Jevon P. Chan holds the position of Senior Engineer at American Bureau of Shipping. The author is responsible for assisting the development of new guidelines for emerging technologies. The author's previous experience includes being a former seafarer and plan approval engineer, alongside completing his PhD at Newcastle University, in 2024 in the field of maritime technology with a research focus of autonomous shipping and human factors.

Autonomous Ships 2024
20th November – 21st November 2024
Copenhagen, Denmark

UNLOCKING KEY ACCEPTANCE CRITERIA FOR SAFE SITUATIONAL AWARENESS SYSTEMS USING SHIP MANOUVRABILLITY CHARACTERISTICS

K B Karolius, **G Løvoll** and **T A Pedersen**, Group Research and Development, DNV, Norway
S A Rørvik and **A L Pavlak**, NTNU, Norway

SUMMARY

The maritime industry is transitioning towards increased autonomous and unmanned operations. To achieve successful commercial realization, safe automated external awareness systems are essential. However, a major safety challenge is identifying suitable performance criteria for timely and accurate estimates of the navigation environment. The external awareness system's performance should enable the interfaced collision and grounding avoidance system to identify and implement timely, safe, and efficient avoidance manoeuvres. These manoeuvres must, as a minimum, be suitable and manageable for the ship's manoeuvring characteristics, making it a governing factor for the awareness system's required performance. This paper presents two methodologies that leverages ship manoeuvring capabilities to identify minimum safe detection distances, an important performance criterion for awareness systems used in ship navigation. One methodology is deterministic, illustrating the influence of governing parameters such as manoeuvrability, encounter heading and speed, whereas the other is probabilistic, enabling probabilistically quantifiable and risk-optimal assessment of detection distances, considering actual operating profiles of the related ships.

NOMENCLATURE

[Symbol]	[Definition] [(Unit)]	[Abbreviation]	[Acronym]
V_{os}	Own-ship speed (kn)	AIS	Automatic Identification System
V_{ts}	Target-ship speed (kn)	ALOS	Adaptive Line of Sight
V_r	Relative speed (kn)	CAGA	Collision and Grounding Avoidance
θ_{os}	Own-ship heading (deg)	CPA	Closest Point of Approach
θ_{ts}	Target-ship heading (deg)	DCPA	Distance between ships at CPA
θ_r	Relative heading (deg)	ESA	External Situational Awareness
α, β	Relative bearing between ships (deg)	ODD	Operational Design Domain
ω	Offset angle to CPA (deg)	PID	Proportional Integral Derivative
S	Distance between ships (nm)	SC	Speed Control
ΔN	Distance in North-direction (nm)	SBMPC	Scenario Based Model Predictive Control
ΔE	Distance in East-direction (nm)	TCPA	Time to Closest Point of Approach
		TC	Track Control

1. INTRODUCTION

Autonomous technologies have advanced rapidly due to the development of software-intensive systems and advanced sensor technologies. The emergence of *Artificial-Intelligence*, supported by an abundance of data from increased connectivity through the *Internet-of-Things*, and the continuous advances in computational power have been the enabling factors. These factors have realised unprecedented capabilities in sensor data-processing and -analysis, allowing for new functionalities in perception technology that were previously unimaginable. As a result, there has been an increase in automated and remotely operated functionalities across various transportation industries, the maritime industry included. Automation in the maritime industry has been driven forward by the desire to reduce costs, increase safety, and address the anticipated future shortage of qualified seafarers.

Adequate awareness of the surrounding navigational environment is a prerequisite to navigate a ship safely and efficiently in accordance with the voyage plan [1]. As the maritime industry transitions towards increased autonomous and unmanned operations, the need for safe automated awareness systems becomes essential. One of the challenges is to identify suitable performance criteria for timely and accurate estimates of the navigation environment. Acceptable performance should enable the collision and grounding avoidance system to identify and implement timely, safe and efficient avoidance manoeuvres, suitable and manageable for the ship's manoeuvring characteristics, making the latter a governing factor for required performance for the awareness system. The IMO's Convention on the International Regulations for Preventing Collisions at Sea (COLREG) [2], set requirements to *whom, how* and *when* to correctly implement collision avoidance manoeuvres. However, the requirements to when manoeuvres should be implemented is given by qualitative terms like: "*in ample time*" and "*apparent manoeuvre*" which are highly ambiguous, open for interpretation and difficult to quantify for an automated system.

The ships' manoeuvrability characteristics is a more quantitative factor governing acceptable performance. Any ship at any speed will have a *point-of-no-return*, represented by the distance and bearing to the target-ship for which the rudder and/or machinery command must be implemented to avoid collision, i.e. *last resort manoeuvre*. This is particularly important seen in relation with Reg. 17 (a) (ii) and b) of the COLREG, where the stand-on ship is required to implement measures to best avoid collision if the give-way ship fails to do so. Before reaching the *point-of-no-return*, adequate awareness is crucial, as well as time available for collision risk assessment and final decision- making. This paper aims to leverage ship manoeuvring capabilities as a foundation to identify minimum safe detection distances, an important performance criterion for awareness systems used in autonomous ship navigation. Two different methodologies will be outlined, one *deterministic*, illustrating how the governing parameters influence the detection distance, such as manoeuvrability, encounter heading and speed. The second method, a *probabilistic* methodology, enables a probabilistic assessment for risk-optimal identification of detection distances considering actual operating profiles of the own-ship and target-ships expected to operate in its *Operational Design Domain* (ODD) [3]. The paper will initially give a brief outline of autonomous navigation and related system modules, followed by the developed methods and results. This is followed with discussions and concluding remarks.

2. AUTONOMOUS NAVIGATION

Before presenting the developed methodologies, a short review of relevant system modules used in automated navigation is necessary to set the problem in a better context, and to outline what is meant by external situational awareness as well as what are relevant parameters that could be measures and invoked with relevant performance criteria. The Autonomous navigation system comprises three main sub-system modules - External Situational Awareness (ESA), Collision and Grounding Avoidance (CAGA) and Track- and Speed-Control (TC/SC), shortly described in the following section. For more information on these systems, reference is made to [4], which also gives a more comprehensive overview of the importance of identifying more detailed performance criteria, a means for providing safety assurance of automated navigation systems. The system modules and dependencies are illustrated in Figure 1 below.

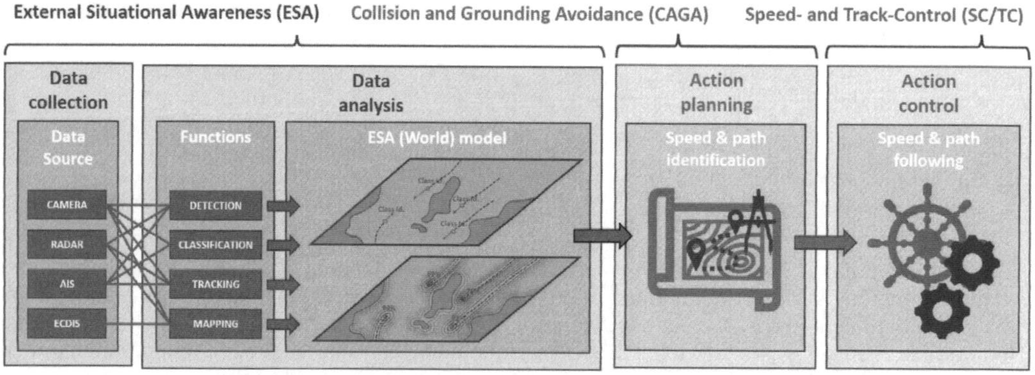

Figure 1: Main system modules and functionalities of automated navigation systems.

The ESA system extracts relevant sensor readings (raw data) from the external navigation environment around the ship, performs data analysis to perceive meaning and context from the data, and aggregates (fuse) the perceived information into a complete model of the navigation environment, often termed a *world model*. The ESA system should have a range of perception functionalities (capabilities), such as *Detection*, *Classification*, *Tracking*, *Mapping*, etc. Each of the functionalities provides estimates or predictions of individual variables that are of interest in the external navigation environment. The estimated variables from each function, can be viewed as distinct layers of information. When combined, they form a comprehensive (synthetic) representation (model) of the external navigation environment. The model quality can be assessed by two factors - the *completeness* of variables, and the *uncertainty* within those variables. Completeness refers to the extent the model represents the real world, characterised by the included information layers and their respective variables. Removing one or more functions from the system, results in reduced layers of information, and decreased completeness. Uncertainty relates to the *correctness* and *accuracy* of these variables, where the correctness involves incomplete or false information, and accuracy relates to how closely the variables match their true state.

The main purpose of the CAGA system is twofold, namely risk assessment and mitigation. The ESA model provides the necessary input to the CAGA system to evaluate and determine the presence of, or potential for, unacceptable collision risks with identified objects in the waterway. This enables to assess and rank potential hazards in need of mitigating actions. Upon assessing the collision risk within the navigational environment, it is essential to mitigate or eliminate any identified hazards. This can be achieved by re-planning a local segment of the overall voyage reference-route, i.e. providing collision avoidance manoeuvres. The manoeuvres must comply with the IMO COLREG, where applicable, and maintain safe underwater keel clearance (grounding avoidance), keeping clear of any restricted zones in the area, while also adhering to the mission-critical parameters, e.g. arrival time, fuel consumption, etc. The Speed- and Track Control systems in combination with the ships propulsion, steering and machinery systems, should enable the ship to execute (follow) the suggested route (control reference) with sufficient precision. The latter is highly dependent on the ship's manoeuvrability characteristics. The route suggestions from the CAGA system have to take into consideration the ships own limitations to manoeuvre, governing its ability to implement the suggested manoeuvres.

From the above short introduction of Autonomous Navigation system modules, particularly the ESA system, it is evident that there are several factors that can be imposed with acceptance criteria, both individually, and in combination. Each function of the ESA system encompasses different features, or layers of information, each with a quality or accuracy level relative to the real world (ground-truth). These parameters are all measurable, allowing to set acceptance criteria to achieve the adequate safety level. The primary challenge lies in determining the acceptable quality-level and achieve required safety-levels, e.g. at what distances from own-ship shall objects be detected, and with what accuracy should their features be replicated? In this paper, the focus have been given the former – At what distance and bearing from own-ship shall target-ships be detected and localised to be safely avoided for various speed and heading combinations? For simplicity, the positioning/tracking accuracy has been disregarded, and it is assumed that when target ships are detected and localised, the accuracy of their position is fully known. Other external environmental disturbances, such as wind, tide and waves have further been disregarded. The accuracy and completeness as well as the environmental influences should all be considered implemented in any future developments on the presented methodologies.

3. METHODOLOGIES

To identify the performance criteria, two different methods will be presented - one *deterministic*, and one *probabilistic*. Both have been developed using the same basic problem formulation and ship simulator framework. The main problem formulation of both methods have been to determine at what relative bearing and distance from own-ship to target-ship the ESA system have to detect the target-ship and estimate its position, enabling implementation of last-resort avoidance manoeuvres to successfully avoid the target-ship with acceptable clearance (DCPA limit), assuming the target-ships, regardless of their actual give-way responsibility in accordance with the IMO COLREG, doesn't take any action. The deterministic method has been implemented using predetermined encounter scenarios, with the main aim at illustrating how ships manoeuvring characteristics, as well as their speed and encounter heading will be governing factors for what is considered acceptable detection distances. The probabilistic method, on the other hand, uses Monte Carlo sampling from representative empirical speed distributions of the ships involved in the encounter scenarios. This enables the utilisation of statistical methods to identify a probabilistically quantifiable acceptance criteria for detection distances. Both methods are presented in the following, although starting with outlining the ship simulation framework used in both. The result from implementing each method is included in their respective sections.

3.1 SHIP SIMULATOR FRAMEWORK

All simulation result presented in this paper have been performed using an in-house (low-fidelity) ship simulator. The simulator have been implemented in the Python programming language [5], comprising individual simulation blocks of the autonomous navigation modules outlined in Figure 1. The ESA simulator includes functionality to simulate the individual ESA functions (variables) with varying degree of added noise. The ESA simulator, in this specific work is highly simplistic, representing the detected position of the target-ship without any noise. The CAGA simulator includes functionality to implement various path-planning algorithms. Specifically for the work outlined in this paper, a simplified version of the *Scenario Based - Model Predictive Control* (SB-MPC) algorithm found in [6] was utilised. The simplification entailed the algorithm to implement full wheel-over last resort actions when exceedance of the DCPA limit was identified to the target-ships. The Speed- and Track control system were implemented using ALOS guidance [7] and a simplified PID controller [8]. The track control system, although not relevant for the simulations presented herein considering simple last resort manoeuvres, would be relevant for the extended methodology suggested under future work. In addition to the elements from Figure 1, as described above, a ship manoeuvrability model from the IEC track-control standard [9], where implemented, representing the physical motion behaviour in 3 degrees-of-freedom for three different ship types of varying size and manoeuvrability characteristics. The model is a simplified model of ship motion established for the objective of testing track control systems. For the reason of ease of use, it has been simplified beyond the normal parameters of a ship. Nonetheless, it is representative of the essential features of ship motion, including:

- the reaction of the propulsion machinery to a command,
- the reaction to a propulsive force,
- straight-line resistance,
- the reaction of the rudder machinery to a command,
- the reaction of the ship to the achieved rudder position, and
- the sideslip motion of the ship during a turn.

The ships specifications and main manoeuvring characteristics are presented in Table 1 below. For more detailed information and derivation of the manoeuvrability model and related coefficients, see the IEC track-control standard.

Table 1. Test ship specifications, hydrodynamic coefficients and main manoeuvre characteristics.

Parameter	**Ship A**	**Ship B**	**Ship C**
Ship type [-]	Ferry	Container ship	Tanker
Length, L [m]	60	250	350
Max speed, V [kn]	30	25	10
Advance [nm]	0.194	0.551	0.666
Transfer [nm]	0.089	0.323	0.349
Tactical diameter [nm]	0.180	0.640	0.760
Final turning diameter [nm]	0.138	0.335	0.481

3.2 DETERMINISTIC METHOD

3.2 a) Method description

The deterministic method has been implemented using predetermined encounter scenarios, with the aim of illustrating how ships manoeuvre characteristics, encounter heading, and speeds govern safe detection distances. Several simulations were performed to identify detection regions around the own-ships. For simplicity, the simulations focused on single target ship encounters. The own-ship heading was set to 0° for all the test ships and scenarios. The own-ship test ships were simulated at three different operating speeds (60%, 80%, 100%) of their maximum speed, as outlined in Table 1. For each own-ship speed, the target-ship was simulated with speeds ranging from 0-20 knots in uniform steps of 0.5 knots, and encounter headings ranging from 0-180 degrees in uniform steps of 1 degree, since the ships are symmetrical and assuming they have the same manoeuvrability characteristics to both sides. The DCPA for the initial condition of the scenario (without implementing any avoidance action) was set to 0 m. Actually, any DCPA below the DCPA limit would be considered a collision scenario, but DCPA = 0 was chosen for the deterministic approach to generate smooth detection distance regions. The initial starting position of the target-ships was calculated based on the initial DCPA of 0, its speed and heading using the trigonometrical relationships illustrated in Figure 2, where the distance S equals $TCPA \cdot V_r$ when $DCPA = 0$. The DCPA limit was set to 60% of the ship length. To determine the detection range (relative bearing and distance) that would enable the ship to successfully avoid the target-ship in line with the specified DCPA limit, a binary search algorithm was employed [10], simulating scenarios with decreasing detection ranges, represented by S in Figure 2, and corresponding wheel-over execution until the minimum detection distance was identified. Both starboard and port manoeuvres were simulated, with the shortest detection distance obtained to safely avoid collision being used, assuming the

CAGA system would be able to identify the optimal side to implement the avoidance action. The complete process behind the deterministic method is outlined in Algorithm 1.

```
Algorithm 1:   Detection ranges, deterministic methodology
set own_ships = [ship_A, ship_B, ship_C]
set target_ship_headings = [0.0, 1.0, 2.0, …, 179, 180] [degrees]
set target_ship_speeds = [0.0, 0.2, 0.4, … ,19.8, 20.0] [knots]
initiate distances = [], bearings = [] (relative between own-ship & target-ship)
for each own-ship do
    set own_ship_speeds = V_OS·[0.6, 0.8, 1.0] [knots]
    set own_ship_heading = 0 [degrees]
    set dcpa_limit = 0.6·L_OS [m]
    for each own_ship_speed do
        for each target_ship_speed do
            for each target_ship_heading do
                while dcpa > dcpa_limit do
                    distance, bearing = search(simulate_scenario(vary_rudder_exec))
                end while
                distances.append(distance), bearings.append(bearing)
            end for
        end for
    end for
end for
```

Algorithm 1. Detection range identification procedure: Deterministic method.

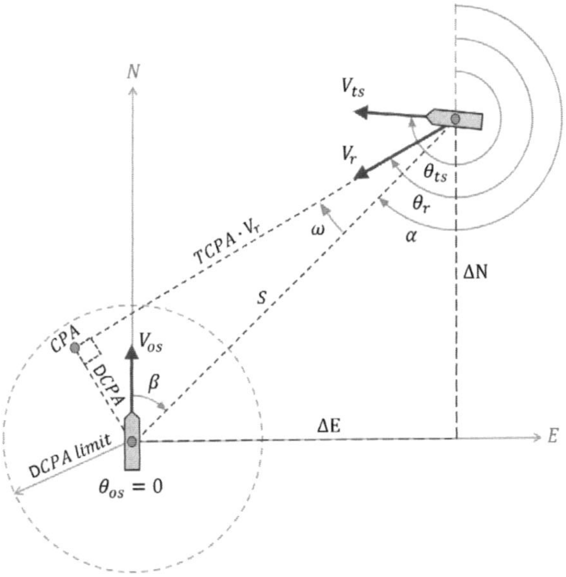

Figure 2: Main parameters of a collision scenario. Figure represents relative velocities.

3.2 b) Method results

The result of implementing the deterministic method is outlined in this section. Figures 3 and 4 illustrate the minimum detection range of the simulated scenarios, where the own-ships A, B, and C are shown in the top, middle, and bottom rows, respectively. The simulated speeds of 60%, 80%, and 100% of the maximum speeds are displayed from left to right. The figures present the identified minimum detection distances for individual combinations of speed and heading for the three own-ships in polar plots. The colour bar on the right-hand side represents the target-ship speeds. Figure 3 has fixed axis, clearly illustrating the differences in distances required for the detection regions depending on the ship type. Ship A, with high manoeuvrability, needs a detection distance forward in the range of 0.35 nautical miles (650 metres) for larger target-ship speeds, whereas Ship B, with less manoeuvrability, needs a detection distance forward in the range of 1.1 nautical miles (2 040 metres) for larger target-ship speeds. Finally, Ship C, with the least manoeuvrability, needs a detection distance forward in the range of 2.5 nautical miles (4 630 metres), almost 600% increase in comparison to Ship A. It can be noted that the difference in detection distance also comes from the choice of DCPA limit being dependent on the ship lengths. It is further observed that the required detection ranges increases with own-ship speed for Ship A, while they decrease with own-ship speed for Ships B and C. The latter is somewhat counterintuitive to what would be expected since the turn circles normally increases in size with increasing speed. This is because, for larger speeds, the own-ship would execute the manoeuvre along the turn-circle path much quicker, regardless of its slightly increase in size, than for smaller speeds, thereby avoiding collision more effectively. This effect is less apparent for Ship A, having a considerable speed, even for the 60% cases. It is further observed that the variation in detection distances with increased own-ship speed is considerable larger for ship C than for Ship A, indicating that speed has less influence on highly manoeuvrable ships. It is further noted that the centroid of all the data points (detection regions) is shifted forward of the ship position along the speed direction, especially apparent for Ship C, which have no datapoint aft of the own-ship for the larger speeds, simply because all target-ships have lower speeds than the own-ship.

Figure 4 illustrates the same as Figure 3 but has scaled axis, enabling a more detailed look at how the minimum detection range varies with speed for the different ships. The centroid shifts observed for Ship C in Figure 3 are now clearly visible for all ships. The colour of the centroids indicates scenarios where the target-vessels have close to zero speed. The centroid shifts for maximum speed are approximately 500 m, 1100 m, and 1300 m along the own-ship speed direction (N-axis) for Ship A, B, and C, respectively. The scaled axes reveal that the detection region extends furthest to each side of the ships at lower own-ship speeds. As speed increases, the region distance becomes more significant in the forward direction for all ships. This is particularly evident for Ship C, where the shape resembles a pointy oval. At the highest speeds for Ship C, the shape is almost like a circle-sector, while Ship A maintains a more circular shape, albeit shifted forward, especially at its highest speed. The pointy shape to each side results from only implementing rudder action and maintaining own-ship speed. When target ships approach from either side, the rudder execution mostly induces transverse motion, which is less effective in avoiding the target vessel compared to when target ships approach from the fore or aft. In the latter case, the transverse motion induced by the rudder action alone is much more effective. This effect is especially apparent for Ship C, which has the smallest operating speed. For target vessels approaching from each side, a speed change could be a more effective strategy, provided the speed change response is sufficiently quick.

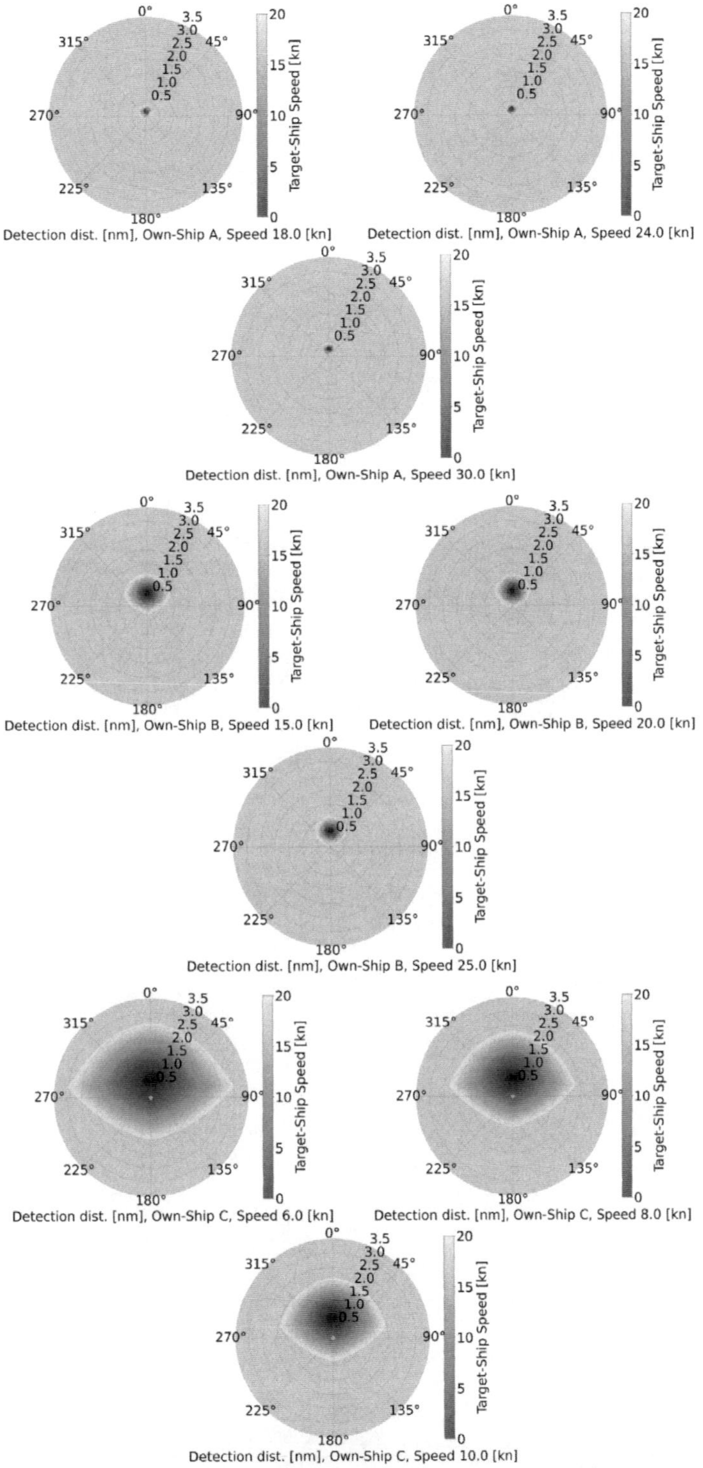

Figure 3: Simulated minimum detection distance using the deterministic method - Fixed axis. Own-ships A, B, and C are shown in the top, middle, and bottom rows, respectively. The simulated speeds of 60%, 80%, and 100% of maximum are displayed from left to right.

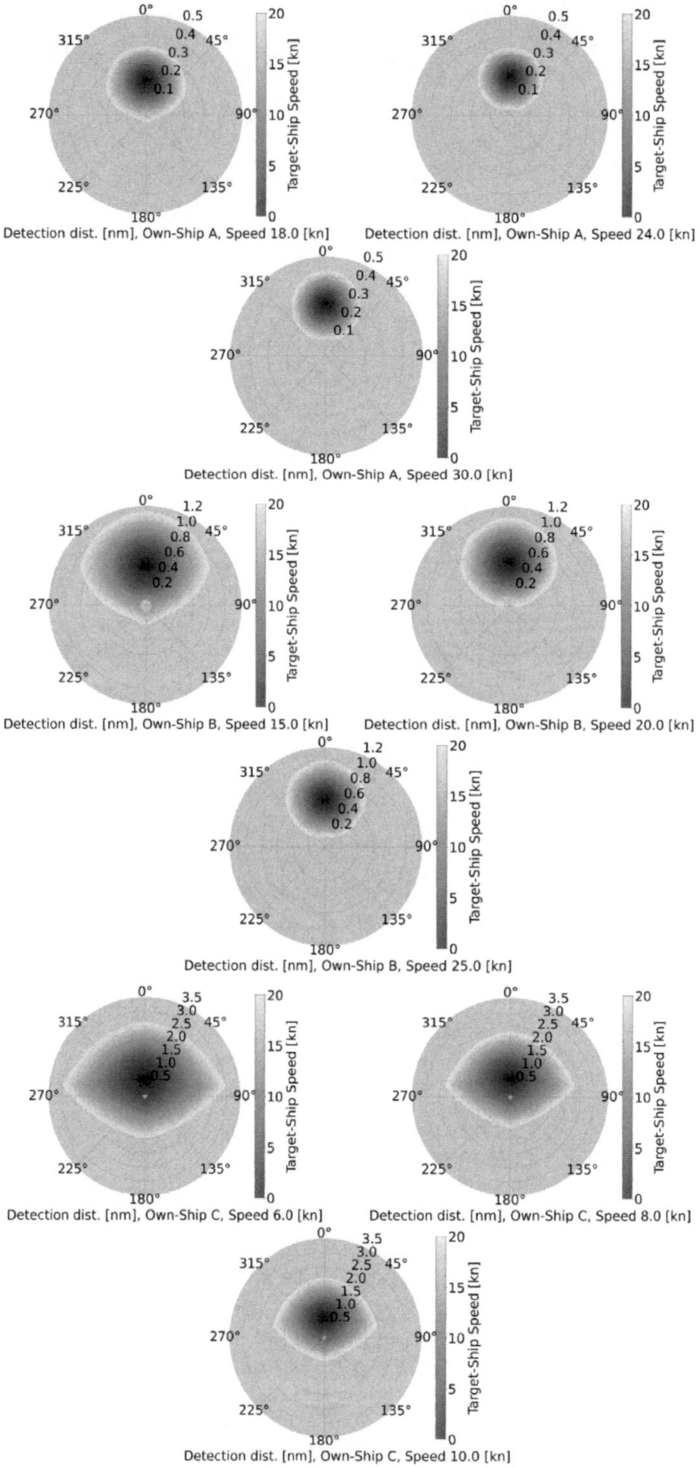

Figure 4: Simulated minimum detection distance using the deterministic method - Scaled axis. The own-ships A, B, and C are shown in the top, middle, and bottom rows, respectively. The simulated speeds of 60%, 80%, and 100% of the max. are displayed from left to right.

3.3 PROBABILISTIC METHOD

3.2 a) Method description

The probabilistic method has been implemented using Monte Carlo sampling from representative speed distributions of the ships involved in the encounter scenarios. The goal is to use statistical arguments to identify a probabilistically quantifiable acceptance criteria for detection distance. For the results presented in the following, a sample size of N = 10.000 scenarios was simulated. Similarly to the deterministic method, the simulations focused on single target ship encounters, with the own-ship heading set to 0° for all test ships in all scenarios. The target-ship headings were sampled randomly between 0 and 180 degrees, and the DCPA limit was set in the same way as in the deterministic approach. The own-ship's speeds was drawn from empirical distributions of ship speed acquired from relevant AIS data of representative ships (matching ship type and size). The reason for using representative ship data is that the methods presented have not been employed on actual ships, but rather on representative ships from the track control standard. For an actual implementation to identify concept-specific performance requirements, the ships' actual expected operating speed distribution should be utilised. The speeds were drawn from the distributions on the range $[20\%, 100\%]$ of the maximum operating speeds, assuming that speeds below 20% represent port and harbour manoeuvring and are irrelevant for normal transit operations where collision avoidance manoeuvres would be relevant. For lower speeds during harbour and port operations, the ships would have insufficient speed and pressure on the rudder to implement any useful last-resort manoeuvring and would rely more on the target-ship to act.

The target-ship speed distributions were also acquired empirically using AIS data but aggregated for all ship types world-wide. For ships intended to operate in a more limited ODD, AIS data for that particular area should be used. The identified speed distributions for the three tested own-ships, as well as the aggregated target-ships, are illustrated in Figure 5. The DCPA achieved for the initial condition of the scenario (without implementing any avoidance action) was, unlike the deterministic method, randomly sampled on the range $[0, DCPA\_lim]$, which enables the sampling procedure to account for all possible breaches of the DCPA limit. The initial starting position of the target ships was calculated based on the initial DCPA, its speed, and heading using the trigonometric relationships illustrated in Figure 2. This is similar to the deterministic method, but includes the CPA offset angle ω, due to $DCPA \neq 0$. To determine the detection region distances that would enable the ship to successfully avoid the target-ship with sufficient DCPA limit, the same binary search algorithm approach as for the deterministic method was employed. Using the results from the simulations enabled a probabilistic detection region to be identified, allowing the aggregated probability of avoiding collision in all expected scenarios (within the expected ODD) to be identified by fitting a circular detection region with its centre offset to the centroid of all the simulated data points. The proportion of points outside the fitted detection region would represent the residual acceptable risk of collision. The complete process behind the probabilistic method is outlined in Algorithm 2.

Algorithm 2: ESE detection ranges, probabilistic methodology

```
set own_ships = [ship_A, ship_B, ship_C]
set N_samples = 10.000
set target_ship_headings = random.between(0, 180, N_samples) [degrees]
set target_ship_speeds = random.choice(target_ship_speed_data, N_samples)
[knots]
set distances = [], bearings = [] (relative between own-ship & target-
ship)
for each own-ship do
    set own_ship_speeds = random.choice(own_ship_speed_data, N_samples)
[knots]
    set own_ship_heading = 0 [degrees]
    set dcpa_limit = 0.6·L_OS [m]
    set dcpa = random.between(0, dcpa_limit, N_samples) [m]
    for i = 1 to N do
        set scene = generate_scenario(i, os_speed, ts_speed, os_heading,
ts_heading)
        while dcpa > dcpa_limit do
            distance, bearing = search(simulate(scenario(vary_rudder_exec)))
        end while
        distances.append(distance), bearings.append(bearing)
    end for
end for
```

```
set N_centroid, E_centroid = calculate_data_centroid(distances, bearings) [m]
set detection_sector = circle(radius = 0, a = N_centroid, b = E_centroid)
while N_data_points ∩ detection_sector < 0.95*N
    radius = search(vary_radius)
end while
```

Algorithm 2. ESA detection range identification procedure: Probabilistic method.

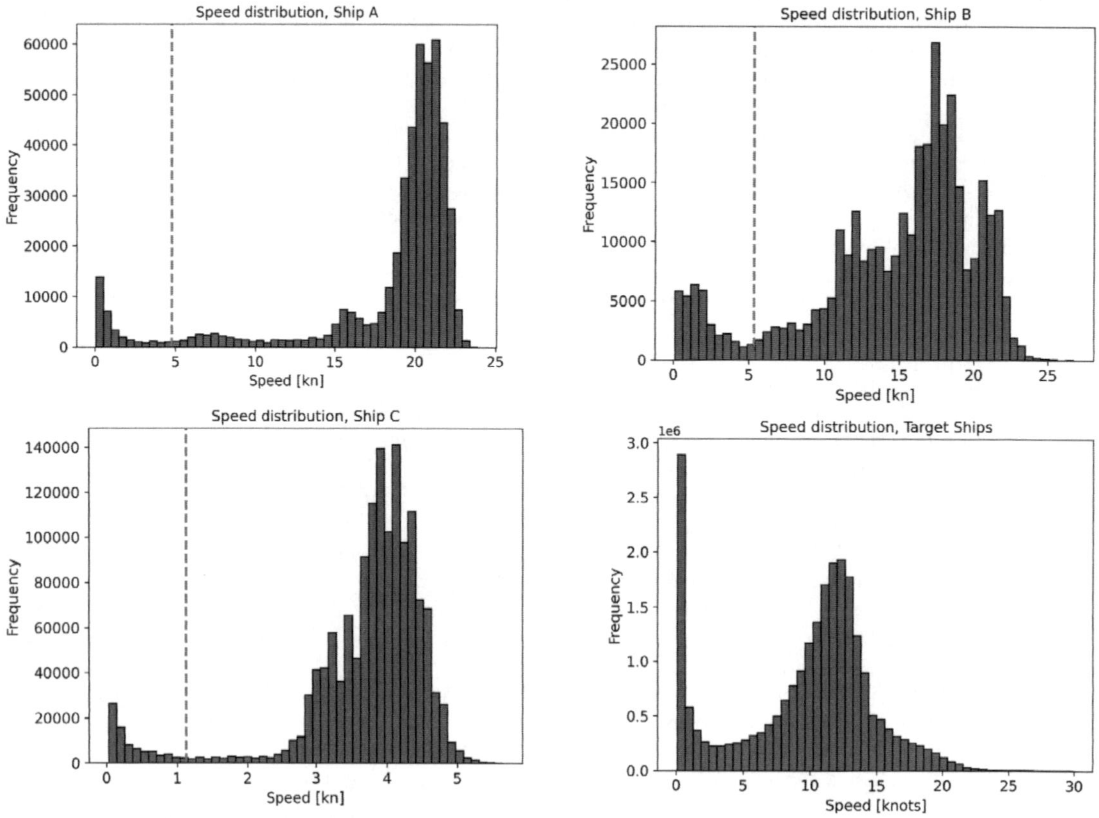

Figure 5: Empirical speed distributions, Own-ships A (472 800 samples), B (352 015 samples), C (141 412 samples) and Target-Ships (24 473 327 samples). Distributions created using AIS data from representative own-ships, and target-ships from world-wide trade. Red stapled line shown in own-ship distributions represents the 20% lower sampling limit.

3.2 b) Method results

The result of implementing the probabilistic method is outlined in this section. The simulated scenarios are illustrated in figure 6 and 7, were the own-ships A, B, and C are displayed from left to right. In contrast to the deterministic result figures, where the colour bar represented the target-ship speed alone, the colour bar now represents the relative speed between the ships. Since the own-ship speeds are now all aggregated in the same figure, using the relative speed enables the illustration of the speed contribution from both ships in the same figure. Figure 6 has fixed axis to illustrate the difference in detection distances for the different manoeuvrability characteristics, similar to what was presented for the deterministic method. In contrast to the deterministic method's smoot detection regions, the probabilistic sampling regime has resulted in a scattering of the data points due to the presented sampling procedure, nonetheless, with a similar shape to the deterministic results. The outer detection distances seems to be in the similar order to those from the deterministic method, however, with a slight increase for all ships for several outliers. This is simply due to the speed distributions for the target-ships allowing for drawing samples in excess of the maximum speed of 20 knots as was used in the deterministic method. From the figure, it is further notable that the largest relative speeds (represented by yellow colour) seems to give the largest detection distances, and mostly in the own-vessel speed direction.

Figure 7, containing the same results, but with scaled axis reveals the shift in data points centroid along the speed direction of own-ship, also for the probabilistic results. The data points centroid shift is calculated to be approximately 0.18 nautical miles (330 m), 0.47 nautical miles (870 m), and 0.60 nautical miles (1100 m) along own-ship speed direction (N-axis) for ship A, B and C respectively. Having all the datapoints and the centroid shifts established, circular detection sectors were fitted to the data, to represent specified proportions (confidence interval) of the data-points falling within the circle. The remaining data-points falling outside of the detection sector represents the acceptable residual risk of collision. The fitted detection regions can be seen in Figure 8, representing 0.95 and 0.99 of the datapoints within the circle, i.e. 95% and 99% confidence that the vessels would be able to avoid a collision if the object is detected before reaching this region. The various detection regions for own-ships and related parameters, for varying confidence intervals are included in Table 2.

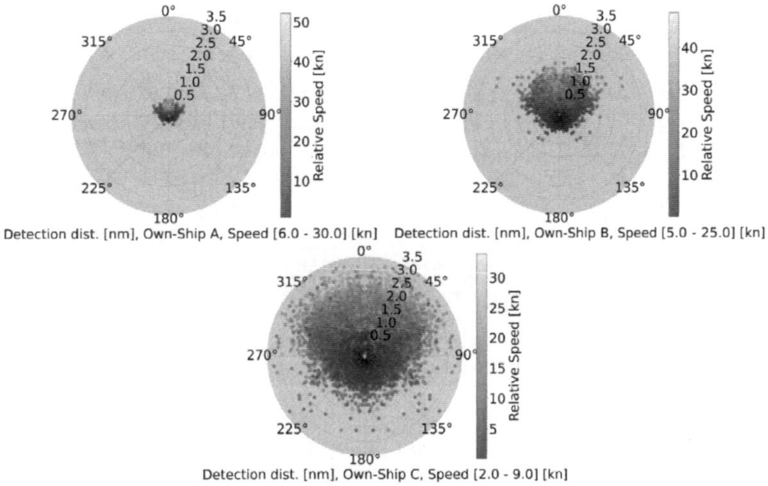

Figure 6: Simulated minimum detection distance using the probabilistic method - Fixed axis. The own-ships A, B, and C are displayed from left to right.

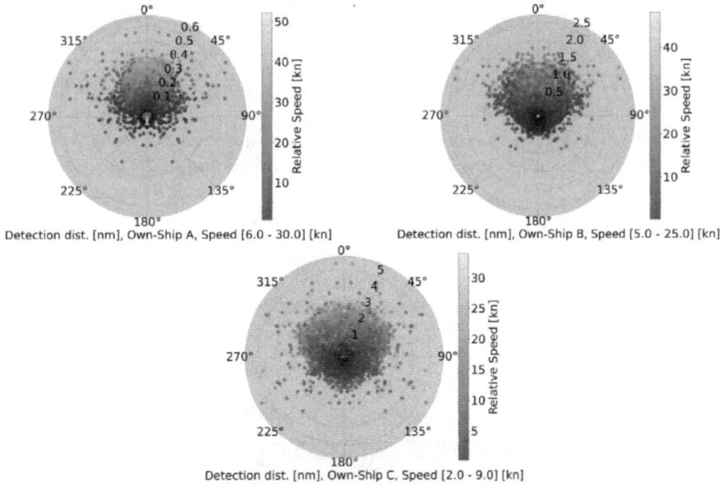

Figure 7: Simulated minimum detection distance using the probabilistic method - Scaled axis. The own-ships A, B, and C are displayed from left to right.

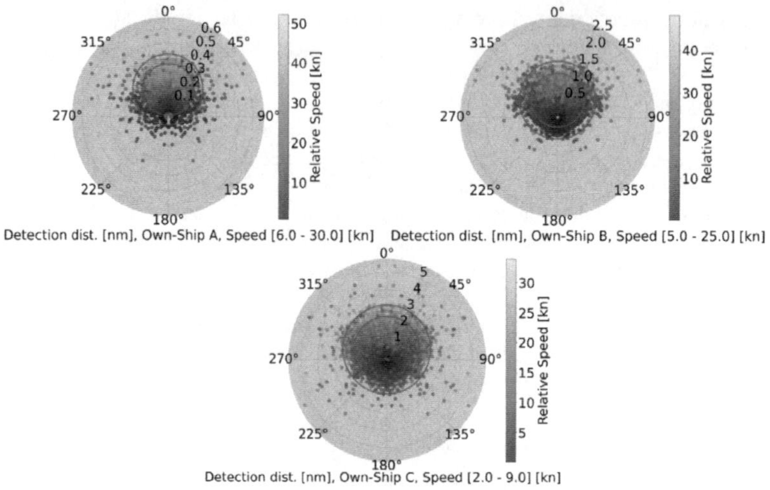

Figure 8: Simulated minimum detection distance using the probabilistic method with scaled axis, including the fitted circular detection section representing 95% (inner-circle) and 99% (outer-circle) confidence in avoiding collision. The own-ships A, B, and C are displayed from left to right.

Table 2. Detection regions, and related parameters, for varying confidence intervals.

	95% Confidence sector		**99% Confidence sector**	
	Radius [nm]	**Offset N [nm]**	**Radius [nm]**	**Offset N [nm]**
Own-Ship A	0.176	0.185	0.241	0.185
Own-Ship B	0.729	0.469	1.058	0.469
Own-Ship C	1.905	0.604	2.539	0.604

5. DISCUSSIONS

In developing and demonstrating the methods outlined in this paper, simplified and generic manoeuvring models have been used from the IEC track-control standard. However, for application on an actual ship concept, a more accurate and representative manoeuvring model should be utilized. This also applies to the speed distributions used in this work. Due to the lack of actual ships, we utilized representative ships that matched those of the three manoeuvring simulation models, when acquiring the AIS data. For the target ships, the distributions are valid but represent worldwide operation. If a more limited ODD is applicable, the data should be gathered for the specific geographic area where the ship is intended to operate. The presented work has not included environmental influences, such as wind, tide, or waves, and how these influence the simulated results. This should be included in any future developments of the methods, where valid data and subsequent distributions may be identified for the wind, tide and wave environment of that particular ODD.

In sampling the target-ship headings, we have limited the sampling range to [0-180] degrees assuming the ships are symmetrical and have equal manoeuvring characteristics to both sides. This symmetry assumption doesn't necessarily hold if we consider ships with asymmetric hulls or single screwed propulsion. This is important to remember if the methods are to be applied to an actual autonomy concept ship towards setting requirements for the onboard ESA system. Another simplification imposed is the exclusion of any uncertainties in the ESA system's ability to estimate the position of the target ships. Even if the target ships are detected, their position estimate and subsequent tracking in time will have some uncertainty related to them. Similarly, other quality measures related to both completeness and uncertainty, as described briefly in the introduction, could be included and simulated when trying to identify the detection regions. This should be considered in future implementation of the method. Finally, multi-ship encounters have not been considered. Often, there is one ship that has a DCPA limit breach, but there may be many other ships (or even landmass) in the vicinity that may affect how the avoidance manoeuvre can and will be implemented to avoid collision safely. This is also an important aspect that should be considered in future implementation of the method. The current simulation setup described in the introduction, with CAGA, Guidance, and Control is, with the above listed limitations, excessive, and several simplifications was used seeing as we only considered last resort manoeuvres in the presented methods. However, the intention is to have many of the above limitations implemented, such as environmental influences and multi-ship encounters. Then the complete end-to-end simulation setup is crucial to assess all elements of the autonomous navigation system.

In the presented methods, the last resort manoeuvre have been limited to a full wheel over (heading change) manoeuvre. In emergency situations, the most efficient manoeuvre for a ship can depend on various factors, including the specific circumstances and the ship's characteristics. Generally, both speed reduction and heading alteration are crucial, but they serve different purposes. A speed reduction is often the first step, but should be implemented already when visibility is reduced, according to Reg. 6 of the COLREG. Reducing speed simply help minimize the impact of a potential collision, give more time to react, and reduces the stopping distance for full-astern manoeuvres, nonetheless reduce the efficiency of the rudder and subsequently reduce the manoeuvrability of most vessels. This is clearly highlighted in the results presented in this paper. Reducing the vessel speed would in most circumstances be implemented when collision is imminent, to reduce the consequence of impact. Changing the ship's heading is a more efficient action but is only useful when there is enough space to navigate around the identified hazard. In any case, a manoeuvre combination of both heading alteration and speed reduction manoeuvres is usually employed. This combination could potentially be considered implemented in the presented methodology but should in any case be seen in relation with the CAGA systems actual last-resort manoeuvre strategy, since this may be different from system-to-system.

6. CONCLUSIONS

This paper have presented two methodologies for identifying detection distances for enabling safe last resort manoeuvres. The initial, deterministic, method was mainly presented to highlight how the various parameters such as manoeuvrability characteristics, speed and heading combinations influence the necessary distances for the detection region. The second method is an example of a probabilistic procedure for identifying a quantifiably risk based safe detection region around the ship, which in turn can be used to set quantifiable performance criteria for the ESA system. The deterministic method implementation results show that detection distances vary significantly based on the ship type and speed. Ship C, a large ship with lower manoeuvrability, requires approximated 6 times the detection distance than the smaller more manoeuvrable Ship A. It was also shown that the required detection ranges increases with own-ship speed for Ship A, while they decrease with own-ship speed for Ships B and C. The centroid of the detection regions shifts forward along the speed direction, and the detection region's shape changes with speed, pointy oval shape for ships with low manoeuvrability, while more circular for ships with highly manoeuvrability.

In conclusion, the deterministic method highlights the importance of considering ship type and speed when determining detection distances, although fails to consider the proportion of time the specific speeds are in operation and would therefore result in unreasonable stringent detection performance criteria. This is solved using the probabilistic method, as actual operating profiles makes the arguments for setting the performance requirements, reflecting the actual ODD of the autonomy concepts. Highly manoeuvrable ships require shorter detection distances, while less manoeuvrable ships need significantly longer distances, clearly highlighting the need for having concept specific requirements on such systems. The detection region's shape and centroid shift also vary with speed, further emphasizing the need for tailored detection strategies for different autonomy concepts. These findings can inform the development of more effective performance requirements for External Situational Awareness systems and help assurance providers assess specific ESA system suitability towards specific autonomy concepts.

7. ACKNOWLEDGEMENT

The work outlined in this paper was produced through the research project SAFE Maritime Autonomous TEchnology (SAFEMATE), supported by the Research Council of Norway under grant number: 327903.

8. REFERENCES

1. THE INTERNATIONAL MARITIME ORGANISATION (IMO), IMO Res. A.893(21), 'Guidelines for voyage planning', adopted on 25 November 1999.

2. THE INTERNATIONAL MARITIME ORGANISATION (IMO), 'Convention on the International Regulations for Preventing Collisions at Sea (COLREG)', adopted 20 October 1972.

3. ADEDJOUMA, M., BOTELLA, B., IBANEZ-GUZMAN, J., MANTISSA, K., PROUM, C M., and SMAOUI, A., 'Defining operational design domain for autonomous systems: A domain-agnostic and

risk-based approach', *19th Annual System of Systems Engineering Conference (SOSE 2024), Tacoma, United States,* 2024.

4. KAROLIUS, K., GRUNDE, L., CHICHOWICZ, J., ENGELHARDTSEN, Ø., and JØRGENSEN, A., 'Navigating the challenges of assuring safe autonomous navigation', *Proceedings of the 2nd International Conference on the Stability and Safety of Ships and Ocean Vehicles, Wuxi China*, 14-18 October 2024.

5. PYTHON SOFTWARE FOUNDATION. 'Python Language Reference', *version 3.12.1 Available at http://www.python.org.*

6. TENGESDAL, T., JOHANSEN, T A.., 'Simulation Framework and Software Environment for Evaluating Automatic Ship Collision Avoidance Algorithms', *IEEE Conference on Control Technology and Applications (CCTA),* August 2023.

7. Automatic Ship Collision Avoidance Algorithms', *IEEE Transactions on Intelligent Transportation Systems, vol. 17, no. 12, pp. 3407–3422,* Dec. 2016.

8. FOSSEN, T I., 'An Adaptive Line-of-sight (ALOS) Guidance Law for Path Following of Aircraft and Marine Craft', *IEEE Transactions on control systems technology,* 17 March 2023.

9. MINORSKY, N., 'Directional stability of automatically steered bodies', *Journal of the American Society for Naval Engineers*, *34(2), pp. 280–309*, 1922.

10. INTERNATIONAL ELECTROTECHNICAL COMMISION (IEC), 'IEC 62065 - Maritime navigation and radiocommunication equipment and systems – Track control systems – Operational and performance requirements, methods of testing and required test results', *ISBN 978-2-8322-4858-4 Edition 2.0 2014-02.*

11. DONALD, K E, 'The Art of Computer Programming. Volume 3: Sorting and searching', 2nd Ed. *Addison-Wesley*, 1998.

9. AUTHORS BIOGRAPHY

Kristian B. Karolius, **Grunde Løvoll** and **Tom Arne Pedersen** currently hold positions as researchers in DNV's strategic research unit, focusing on remote-controlled and autonomous ships, systems, and related technologies. Their main research focus is on developing assurance standards and requirements, tools, and related infrastructure for testing autonomous navigation systems. This includes work on situational awareness, and collision avoidance systems and related components.

Adrian Langmo Pavlak and **Simen Andreas Rørvik** are both undertaking a MSc degree in Cybernetics and Robotics at NTNU, currently finalizing their fifth and final year of the degree. They have both been employed as research interns in DNV's strategic research unit focusing on the development of simulation methodologies for testing autonomous ship systems.

BUSINESS ECOSYSTEMS IN AUTONOMOUS SHIPPING: A REVIEW-BASED ACTOR MAPPING

K Kloch, Aalborg University, Denmark

SUMMARY

Departing from the concept of business ecosystem within management research, this paper represents the preliminary findings of a comprehensive literature review of autonomous shipping business ecosystem. Based on the review, it was found that the alignment structure between ecosystem actors needed to be mapped as a natural first step in dissecting the research landscape and highlighting the potentially transformative nature of autonomous ships. Findings indicate that changes in the roles of actors could be expected, with possibly new roles emerging and old roles disappearing. Concludingly, the study highlights the need for the business ecosystem concept to be used within autonomous ship research as the ships cannot be seen in isolation due their potentially disruptive impact on the context in which they are deployed – which in turn may make them unfeasible if the associated ecosystem is not aligned correctly.

1. INTRODUCTION

While "economy-of-scale" has driven down transportation costs and emissions by constructing increasingly larger vessels, the paradigm has also strained existing infrastructure and reduced overall supply chain resilience [1]. Notable incidents, such as the Ever-Given grounding in the Suez Canal, illustrate its vulnerabilities with cascading disturbances across supply chains [2]. As only few ports can make the large investments required to accommodate the large vessels, the supply of available infrastructure has decreased, resulting in concentration of routes and cargo [3]. Global trade is hinged on the ability of local economies to offer feasible and reliable ways of moving goods over long distances, noticeably through the effective use of sea ports, which enables trade opportunities with economies outside its immediate regional context [4]. Consequently, the reduction of available ports will slowly lead to a subsequent disintegration of the world trade, with peripheral economies finding it increasingly difficult to participate in the global shipping market (see e.g. the cases of Indonesia [5] and of Africa [6]).

Autonomous ships are seen as a potential solution to these challenges, offering substantial financial and environmental benefits which are only expected to increase as the ships become smaller and more standardised [7], enhancing resilience and promoting more diverse use of trade hubs. Nevertheless, one must transcend traditional frameworks to fully unlock the potential of autonomous ships and the associated new business models that can be enabled [8]. Current studies on autonomous ships, however, emphasize the economic benefits for the ship operators in isolation yet tend to overlook potential impacts to other supply chain actors [9]. Consequently, there is a need to investigate autonomous ships in the broader business ecosystem they inhabit, including how the various actors in the ecosystem may need to change for the value proposition to be delivered.

This paper represents the first findings of a comprehensive literature review of business ecosystems within autonomous shipping. As the review developed, it became necessary to dedicate an individual investigation of the alignment structures between actors within the ecosystem in order to pay sufficient attention to their transformation and realignment. The remainder of the document is thus organised as follows: First, a brief introduction to the business ecosystem concept is presented. Second, the research methodology is provided. Third, the analysis of the ecosystem actors as found in the literature are described and mapped. Fourth, the findings are discussed against real-life developments. Fifth, a conclusion is given that summarizes the findings of the study, including the limitations and natural next steps of the research.

2. BUSINESS ECOSYSTEM AS A CONCEPT

In their literature review on the concept of business models, Zott et al. [10] find that the term does not have a clearcut definition. Although the Business Model Canvas framework [11] is widely used, the framework has been critiqued for not being able to capture how the company itself connects with the wider context of its operation (see e.g. [12] [13] [14]). Nielsen et al. [15] suggest that business models are designed and implemented in specific environments which at the same time both influence and are influenced by the business models themselves. In this regard, *business ecosystem* has proven to be a welcomed concept. The term was first coined by Moore [16] to describe market competition through the analogy of predators and prey as a necessary extension to Porter's [17] concept of value chains. Today, thinking in ecosystems have become an increasingly popular approach in business and management research (see e.g., [18] [19] [20]).

While Moore's original conceptualisation saw ecosystems as *affiliations*, newer research offers a complementary approach by viewing ecosystems as *structures* [18]. Essentially, whereas *ecosystem-as-affiliation* sees ecosystems as communities of actors defined by partner numbers, network density, and actor centrality etc.; *ecosystem-as-structure* sees ecosystems as configurations of activities defined by a focal value proposition towards which the actors collaborate [18]. Inspired by Tsvetkova & Hellstrom [8], this study uses the latter approach, defining ecosystems as *"the alignment structure of the multilateral set of partners that need to interact in order for a focal value proposition to materialize"* [18, p. 40]. The study thus investigate the shifting relationships and interactions between ecosystem actors in order to deliver the focal value proposition of autonomous shipping. For the purposes of this study, a wider interpretation of business ecosystem research of autonomous shipping has been taken, including research that addresses the underlying components ("alignment structure", "multilateral", "set of partners" and "focal value proposition" [18]) without explicitly referring to ecosystem as a concept.

3. METHODOLOGY

In their extensive analysis of search systems, Gusenbauer & Haddaway [21] find that only half of the reviewed search systems can be recommended for evidence syntheses research. Following an initial test query on the recommended platforms, Bielefeld Academic Search Engine (BASE), Scopus and Transport Research International Documentation (TRID) were chosen as the search engines for this study. Another determinant in finding the specific literature covering a precise field of study is search queries [22]. Consequently, for the purpose of this study, the final search query (see figure 1) was refined over multiple iterations, including various combinations of words related to autonomous shipping and business ecosystem.

(**"autonomous"** OR **"unmanned"** OR **"automation"**) AND
(**"ship"** OR **"barge"** OR **"ferry"** OR **"maritime"** OR **"shipping"**) AND
(**"ecosystem"** OR **"business model"** OR **"transport system"** OR **"transport network"** OR **"supply chain"**)

Figure 1: Final search query

The study has been guided by the search results, filtering off irrelevant material and including additional literature based on snowballing of references and citations. This was done to ensure that the final corpus contained all the essential research within the field of enquiry. The search journey can be explained as follows: First, literature search was conducted on the BASE, Scopus and TRID search engines. Second, the corpus was filtered for relevant titles, keywords, types, abstracts etc. Third, both forward and backward citation searches were conducted on each of the remaining documents to uncover additional sources of relevance not within the search query itself. Fourth, the remaining texts were read in full to inspect content and identify ecosystem actors, with irrelevant and out-of-scope material filtered out. The final corpus thus contained 43 documents.

Given that the analysis is grounded in a review of the available research landscape, it may be that some key actors are absent from this study due to their absence within the research landscape. This does however not mean that they are irrelevant for the autonomous ship business ecosystem. Another iteration of this study should look beyond the present academic research landscape and include even more actors, concentrating not only on the maritime ecosystem but the transport ecosystem at large.

4. ANALYSIS

Kim et al. [23] find that especially ship design and shipbuilding will be impacted by autonomous ships. Tsvetkova & Hellstrom [8] echoes this by arguing that the focal value proposition within the ecosystem may change by moving the sector into digital servitization, which in turn may result in a reshuffling of the leading roles and actors within the ecosystem. Additionally, Tsvetkova et al. [24] highlight that as the value proposition of autonomous ships extend beyond the traditional ecosystem, the roles are foreseen to shuffle even further with the introduction of new actors. Domenighini [25] echoes this, noting that not only may multiple roles become significantly different, they may even become entirely obsolete. Similarly, as Aiello et al. [37] note, many shipping companies have already integrated their supply chains today and are thus also engaged in port operations and land-based transportation, especially as a result of the containerisation of transport goods. A similar trend may likewise happen with the "autonomisation" of shipping.

Sandvik et al. [26] find that ambidextrous and ambivalent roles will appear throughout the ecosystem transformation, creating tensions between making novel partnerships in the new ecosystem and satisfying existing partnerships in the old ecosystem. This is paradoxical both because ecosystem actors are stretched between maintaining their positions in the traditional ecosystem versus taking on a new position in the emerging ecosystem and because even when they assume the same role in both ecosystems, they are participating in competing ecosystems [26]. While this may indicate that actor roles are not as explicit and fixed as described in this study, a generalisable mapping of the alignment structure in the autonomous shipping ecosystem can be seen in figure 2.

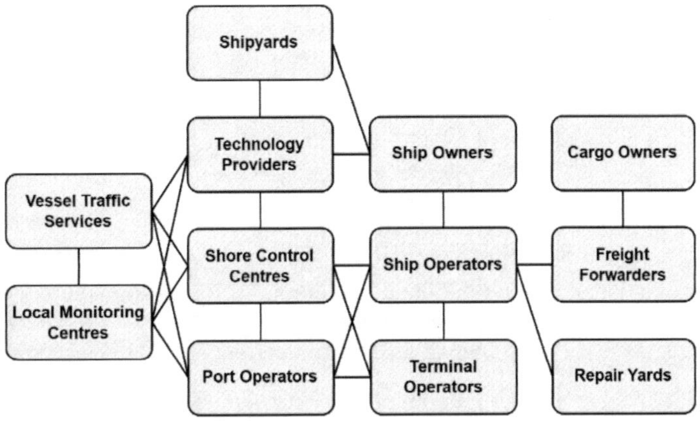

Figure 2: Ecosystem alignment structure following a non-disruptive implementation of autonomous ships

4.1 TECHNOLOGY PROVIDER AND SHIPYARD DRIVEN ECOSYSTEM ALIGNMENT STRUCTURE

The first noticeable shift in the alignment structure is the opportunities arising for shipyards and technology providers – especially the latter. Munim [27] argues that the technology providers in traditional industries such as the shipping industry work to transform the very industry they inhabit towards higher levels of technological maturity, which in turn change the business models and the alignment structures in the ecosystem. Autonomous ships may enable new services and revenue streams by moving from a transport-centred, product-based contract to a technology-centred, service-based contract, which in turn open the door for restructuring of asset ownership and management [27]. In the case of shipbuilding, Sandvik et al. [26] find that with autonomous ships and their wider transport solution(s), the contractual chain may reverse: traditionally, shipowners would order their assets through a shipyard, who would gather tenders from various sub-supplying technology providers. However, autonomous ships may instead be ordered directly from a technology provider, who in turn would gather tenders from sub-supplying shipyards [26].

Domenighini [25] proposes that the manufacturer of the ship may get the possibility to offer both ownership and vessel-sharing services to their clients, with accompanied after-sale services such as maintenance or updates. This same service-based business model configuration could also be applied to the interrelationships between technology providers themselves and their connection to the shipyards and other ecosystem actors [25]. However, Ghaderi [9] finds that shipyards themselves may not have the necessary skills to develop autonomous ships in-house, while trying to acquire these multifaceted and complex skills may usher a gold-rush for talent across the ecosystem with only the larger firms with the breadth and depth to acquire and/or develop the skills surviving. Consequently, shipyards may be positioned as central integrators within a complex network of technology providers who provide individual modules or services for the larger system of systems [9]. Domenighini [25] echoes this by proposing that additional value propositions could be found at various micro-levels between a subgroup of interdependent technology providers within the ecosystem.

Tsvetkova & Hellstrom [8] inform us that the full value proposition of autonomous ships come via entirely new business models which simultaneously curb potential liability issues that may present barriers to operation. Depending on the ship owner and operator, these liabilities may alter, which in turn may influence insurance opportunities and ultimately the ships' overall feasibility [8]. Similarly, Domenighini [28] notes that if only a single actor assumes the entire responsibility because of various legal and contractual bottlenecks, the interests in investing and operating autonomous ships may be discouraged. The ship operators may thus get trapped with liabilities claim arising from cargo interest parties without themselves having claims against the technology provider who delivers the autonomy capabilities onboard [28]. Consequently, the technology provider may be the only actor capable of managing this risk, if entirely new actors are not invented from a risk limitation perspective [28].

Figure 3 shows this realignment of the business ecosystem from the viewpoint of shipyards and technology providers

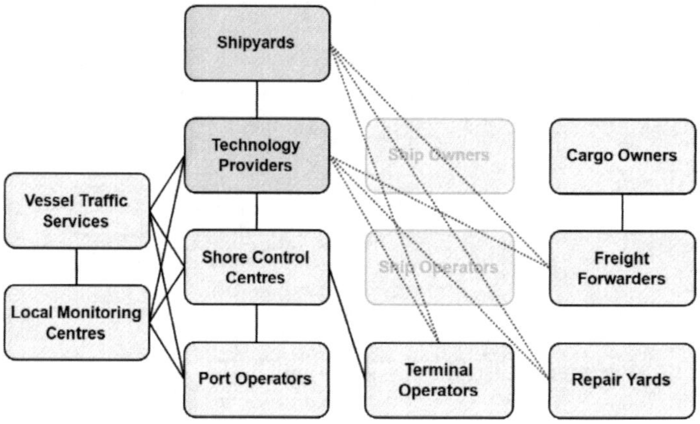

Figure 3: Ecosystem alignment structure with shipyards and technology providers as focal actors. Blue lines indicates new interrelations.

4.2 PORT AND TERMINAL DRIVEN ECOSYSTEM ALIGNMENT STRUCTURE

The second noticeable shift in the alignment structure is the cruciality of ports and terminal operators. Rodseth [7] argues that autonomous ships will depend on port infrastructure, including the need for automated port-based activities. Katsivela [29] echoes this, mentioning that autonomous ships are only one component that will shape the future of shipping, with the introduction of advanced technology also a must in other parts of the transport system. Similarly, multiple authors argue that autonomous ships need an array of complementary innovations around them, exemplified by ports becoming responsible for many activities previously performed *en route* (see e.g. [7] [9] [24]). Furthermore, Tsvetkova & Hellstrom [8] find that ports need to adjust their activities and infrastructure not only to accommodate autonomous ships, but also to create further value through seamless operations in the ports themselves. Autonomous ships will likely spend longer time in port undertaking these new activities, which may in turn lead to deteriorating value for port operators whose business model currently is built around maximising throughput of vessels and cargo [8]. However, cf. Domenighini [25], ports could transform to include the operation of autonomous ships within and in extension of the port's jurisdiction. The ports may thus increasingly find themselves as an even more central actor for the success of autonomous ships, potentially with the investing in and operation of autonomous ship as a line of business in their portfolio [25].

Ghaderi [30] brings our attention to the fact that traditionally, a ship is deployed based on the alignment between the cargo type and the port operations requirements, showing that ports have always been a crucial component in shipping. Ports also play a major role in the competitiveness of ship solutions based on the service quality and cost effectiveness of the ports themselves, with today's seafarers involved with optimising the port stay for quick turnaround [30]. In his later work, Ghaderi [9] continues that the port sector may potentially experience resulting job creation via new service offerings, which may increase the revenue stream of the ports and terminals. From an operational view, however, it may be difficult for ports to undertake such responsibilities in terms of skill, equipment and capacity, particularly for small to medium sized ports where autonomous ships are seen to be most useful [30]. Getting the operation right may in turn result in wider economic benefits for the region surrounding the small and medium sized ports, potentially making actors such as communities, local governments and industries located herein interested in participating in novel investment and operational models for the port business [9]. Kurt & Aymelek [31] echo this, identifying the interaction between autonomous ships and ports as a complex issue, with ensuring continuous communication and collaboration between ports and autonomous ships a noticeable challenge. Sceptics argue that increased berth occupancy and port traffic could decrease efficiency, but the shift to autonomous ships also includes potentially smaller vessel sizes and enhanced control over supply chains, which in turn may offer new turnaround opportunities [31]. That is, ports that invest in infrastructure for autonomous ships and make themselves "autonomous ship friendly" may gain a competitive edge in the future. More recently, Kurt & Aymelek [32] also find that there will be a need for autonomous ship-port integration in the many bureaucratic and commercial processes to enable viable autonomous shipping.

Krause et al. [33] highlights that the high capital expenses (CAPEX) involved with acquiring the necessary infrastructure, equipment and processes required for autonomous ships provide high financial risks. This is especially difficult for small and medium sized ports, in which smarter investment strategies must be enabled so that other ecosystem actors can engage in joint investments of a particular autonomous shipping business case [33]. Nesheim et al. [34] continues on this topic, highlighting the significant asymmetry in investment needs for autonomous shipping: Ports must make substantial up-front investments with limited flexibility and face considerable financial risks if contractual agreements are not established to guarantee returns. This is further complicated by the disjointed profit margins among actors in the ecosystem. Thus, the authors suggest business model innovations, including long-term contracts, shared ownership of equipment, maximizing equipment utilization, and rethinking the transactional model of ports [34]. Furthermore, Psaraftis et al. [35] highlights that ports do not abide by standardised cost or price structures, with noticeable range between the consequent performances of ports. This makes it difficult for the ecosystem to align its collaborative structure – especially where prices are not publicly available a priori [35].

Figure 4 shows this realignment of the business ecosystem from the viewpoint of the ports and terminals.

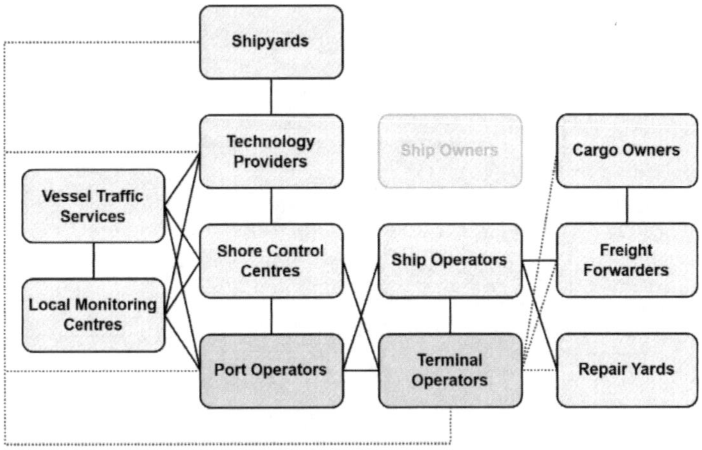

Figure 4: Ecosystem alignment structure with ports and terminals as focal actors.
Blue lines indicates new interrelations.

4.3 CARGO OWNER DRIVEN ECOSYSTEM ALIGNMENT STRUCTURE

The third noticeable shift in the alignment structure is the potential for cargo owners to have direct control over their supply chains via autonomous shipping. Reiterating that traditionally, a ship is deployed based on the alignment between the cargo type and the port capabilities [30], it becomes obvious that the cargo type is fundamental for the feasibility of autonomous ships and the business ecosystem(s) that are aligned around them. For these ecosystems to thrive, however, Rodseth et al. [36] argue for more standardised cargo units that can enable shipment to a broader range of possible ports and terminals in order to avoid route centralisation and increase resilience. Furthermore, Ghaderi [30] argue that cargo owners could own autonomous ships, which would offer a remarkable cost saving opportunity in the overall prize of producing and delivering goods by potentially cutting out various middlemen. Further still, Kim et al. [23] argue that the future of transportation will be a timely service orchestration that allows cargo owners to instantaneously customise shipments, which in turn creates options for just-in-time transport and a broad pool of data insights that could be converted to revenue streams [23]. This idea is again echoed by Ghaderi [9], who claims that network capacity could be leveraged optimally and new as-a-service business offerings may emerge in which provision of cargo transportation solution is more focused on the needs of cargo owners as opposed to the type and capability of ships and ports and their availability in the moment. The business case for cargo customers is also highlighted by Tsvetkova & Hellstrom [8], who see autonomous ships as an opportunity for cargo customers to realign the ecosystem structure such that they get unprecedented control on how their goods are produced and transported by creating fully automated and transparent supply chains. However, the cargo owners are not typically involved in the investment decision unless it is a vertically integrated logistics chain, which means that complementary business models or services may need to be created or that the alignment structure in the ecosystem must transform in a way that allows for such investments [8].

Ahn et al. [37] mention the potential of autonomous ships in the context of a nearshoring tendency in industrial manufacturing, where an increasing focus on more sustainable and resilient supply chains result in industries looking to substitute "offshoring" with "onshoring" or "nearshoring" whereby production sites are moved back to the countries of the consumers or in near vicinity thereto. As manufacturing plants themselves also become increasingly automated and autonomous, active labour becomes less involved in the manufacturing process itself, which makes the previous advantages of "offshoring" manufacturing plants of labour-intensive industries to regions where lower manufacturing costs can make up for higher transport costs less relevant [37]. The possibilities enabled by autonomous ships seem only to accelerate this new trend, with the purchasing and deploying autonomous ships merely an extension of autonomous manufacturing in which production and transportation is seamlessly integrated in a larger autonomous ecosystem [37]. This is echoed by Makkonen et al. when arguing that autonomous ships can enable digital corridors specifically and digital trade more generally, with the ships themselves used as floating stock in just-in-time supply chains [38]. This in turn may decrease the need for warehousing and land-based storage space significantly, with floating stock automatically redirected to any port based on port congestion, supply-and-demand etc.

Figure 5 shows this realignment of the business ecosystem from the viewpoint of the cargo owner.

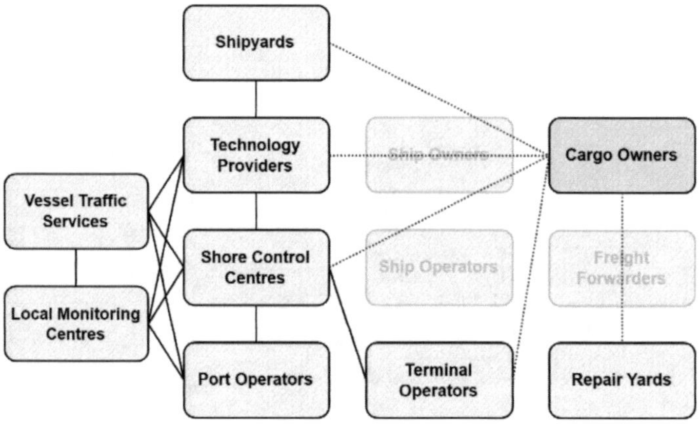

Figure 5: Ecosystem alignment structure with cargo owners as focal actors.
Blue lines indicates new interrelations.

5. DISCUSSION

The actor mapping of this study was based on an attentive literature search and review strategy to also encompass literature that may address ecosystem components without explicitly referring to the business ecosystem concept. This was followed by mapping out both present and resulting actor roles within the ecosystem under the introduction of autonomous ships.

Shipyards and technology providers may see their role altered with the rise of autonomous ships, with especially the latter gaining a crucial position in enabling the value proposition of the ecosystem. Traditionally, shipowners would order ships through shipyards, but with autonomous ships, it may instead be the technology providers who manage the designing, building and outfitting processes. However, this argument could be said to exist for the present shipbuilding market, which does not see the shipyards and technology providers as the main owners and operators of ships – especially not in the context of existing contractual and extracontractual liabilities related to carrying cargo by sea. Given that shipyards and technology providers may prefer to remain non-liable in this regard, it is debatable whether they are interested in integrating autonomous shipping into a more central part of their service offerings. However, given the potential multiplicitous and interdependent complexity of future waterborne IT systems, bestowing ultimate responsibility and/or liability on a single actor may become a futile endeavour in itself. Instead, the industry may find that new, intermediate entities will be created. MASSterly, the joint venture between Kongsberg (technology provider) and Wilhelmsen (ship operator) may be an example of such entity [39]. This presents a highly relevant and significantly understudied area of investigation, especially considering the focus of academic literature on the potential of technology providers to bypass the traditional maritime actors such as ship owners and operators while simultaneously stimulating a monopolising effect within the market of technology provision. Thus, one should ask if this is even of interest to the shipyards and technology providers.

Ports and terminals must upgrade their infrastructure and implement advanced technologies to accommodate autonomous ships, while also assuming responsibilities previously managed by the crews onboard. This leaves them with high financial risks, especially in small to medium-sized ports, necessitating collaborative business model configurations within the ecosystem. Similar to above, it remains uncertain whether ports and terminals would be interested in taken on these new responsibilities – and if they would even be able to. Oppositely, ports and terminals that invest in autonomous ship-friendly infrastructure may gain a competitive edge in the future, potentially even including the ownership and operation of autonomous ships within their business portfolio. The extensive inland waterway network of BCTN is an example of a relatively small inland terminal operator which has transformed into a more holistic logistics provider that offers terminal handling, inland waterway transportation and last-mile trucking [40], with significant cargo owners in their portfolio such as e.g. the clothing company Nike [41]. Furthermore, ports are increasingly seen developing extensive partnerships with other ports in order to enable and coordinate logistics services, exemplified by the merging of separate ports into a single entity (e.g. the Antwerp-Bruges merger [42]) as well as extensive partnership building between independent ports themselves (e.g. the Antwerp-Bruges and Duisburg collaboration [43]). While this may not be new developments, it provides examples of the willingness of ports to take on new roles within supply chains in general. As ports and terminals become more active within the business ecosystem, enabling both autonomous ships in particular and autonomous logistics in general, it may not be farfetched to think that ports will be interested in taking a more central part of the cargo transport itself. Ports and terminals may also want to run interterminal services with autonomous vessels within their own premises in order to optimise throughput by avoiding terminal and fairway congestion, seeing the port as one big warehouse in which units are shifted around based on integrated data exchange (see e.g. the MAGPIE project in the Port of Rotterdam [44]). However, similar to the above, understanding the potential business interest of ports and terminals in the active investment and management of autonomous ships is not only an understudied area of research, it is also fundamental to the development of autonomous shipping as a feasible solution in reality.

Cargo owners stand to gain an unprecedented control over their supply chains, potentially allowing them to bypass traditional intermediaries. Autonomous ships may enable custom shipping services and real-time transport optimisation, which could give cargo owners more flexibility and transparency in their production and supply chains. The ships could also serve as "floating stock" in just-in-time supply chains that minimize the need for warehousing, while also allowing automatic redirection based on factors like port congestion and enabling a much greener supply chain in general. While not explicitly operated by a cargo owner, the BCTN Meerhout terminal is tailored to favour the Nike European Logistics Campus with inland waterway connections to both the major oversea port of Rotterdam [45]. Tailoring a logistics service like this around the specific needs of cargo owners may be seen as the first logical step towards the cargo owners themselves taking over the supply chain orchestration in the light of autonomous ship services. The chemical company Yara, specialising in mineral fertilisers, may be an example of the subsequent step in this development, with the Yara Birkeland specifically deployed to transport fertilizer between the Yara production plant to the regional export port in Norway – and from there onward into the larger supply chain [45]. Other actors such as IKEA and Amazon are already known for buying own containers and chartering own vessels to ensure control of their supply chains which are increasingly subjected to bottlenecks, disruptions and delays [46] [47]. As industries are expected to move towards on- and/or nearshoring of manufacturing [37], cargo owners might even own autonomous ships themselves, seamlessly integrating production and transportation into an encompassing autonomous industrial ecosystem. At present, however, this is an understudied area of research in general, especially through the lens of autonomous shipping business ecosystems.

6. CONCLUSION

The aim of the paper is to introduce the shifting roles, relationships and interactions between a multilateral set of actors in the maritime ecosystem to deliver the value proposition of autonomous shipping. The findings show that multiple actors and their interrelations may be affected; especially shipyards, technology providers, ports, terminals and cargo owners could become focal points in the realigning of the ecosystem structures.

While the study shows a novel mapping of the interdependences of ecosystem actors and how these shifts with the introduction of disruptive technologies, the analysis has limitations. Although the literature search aimed broadly, it may still have missed crucial research existing outside the used search engines and queries. Also, some relevant actors may be absent in the specific corpus and general research on autonomous ships, which makes them absent also in this analysis.

The study is an example of the advantages of using the business ecosystems concept to analyse autonomous shipping, making explicit the alignment structures of business ecosystems that must realign in order to make viable business case(s) for autonomous shipping. As shipping services are deployed within a complex context of consumers, producers and a wide range of intermediary actors, autonomous shipping must necessarily be studied in the broader perspectives.

Finally, the study presents the preliminary findings of a comprehensive literature review on autonomous shipping business ecosystems, wherein the alignment structure of the multilateral set of actors within the ecosystem was a necessary first step in dissecting the research and highlighting the potentially transformative nature of autonomous ships. Consequently, further research on the themes in autonomous shipping though the lens of business ecosystems is forthcoming.

7. REFERENCES

1. OECD; ITF, 'The Impact of Mega-Ships', OECD; ITF, 2015.

2. CAMACHO, F J., 'Ship that destroyed the Baltimore bridge was big. It wasn't the biggest', *E&E News by Politco*, 2024.

3. YEE, V., and GLANZ, J., 'How One of the World's Biggest Ships Jammed the Suez Canal', *The New York Times*, 2021.

4. SAXON, S., and STONE, M., 'Container shipping: The next 50 years', *McKinsey & Co. - Travel, Transport & Logistics*, 2017.

5. MUNIM, Z H., and SCHRAMM, H., 'The impacts of port infrastructure and logistics performance on economic growth: the meditating role of seaborne trade', *Journal of Shipping and Trade, vol. 3, no. 1*, pp. 1-19, 2018.

6. HARRIS, E., SAIDIN, O K., SIRAIT, N N., and KABAN, M., 'Strenghtening National Logistics Ecosystem to Increase Indonesia Competitiveness in International Trade', *Proceedings of the Second International Conference on Public Policy, Social Computing and Development (ICOPOSDEV 2021), vol. 642*, pp. 248-253, 2022.

7. OLAPOJU, O M., 'Autonomous ships, port operations, and the challenges of African ports', *Maritime Technology and Research, vol. 5, no. 1*, p. 260194, 2023.

8. RODSETH, O J., 'Assessing business cases for autonomous and unmanned ships', in *Technology and Science for the Ships of the Future. Proceedings of NAV 2018, Trieste, Italy*, 2018.

9. TSVETKOVA, A., and HELLSTROM, M., 'Creating value through autonomous shipping: an ecosystem perspective', *Maritime Economics & Logistics, vol. 24, no. 1*, pp. 255-277, 2022.

10. GHADERI, H., 'Wider implications of autonomous vessels for the maritime industry: Mapping the unprecedented challenges', *Advances in Transport Policy and Planning,, vol. 5*, p. 263–289, 2020.

11. ZOTT, C., AMIT, R., and MASSA, L., 'The business model: Recent developments and future research', *Journal of Management, vol. 37, no. 4*, pp. 1019-1042, 2011.

12. OSTERWALDER, A., and PIGNEUR, Y., 'Business Model Generation: A Handbook for Visionaries, Game Changers and Challengers', *New York, NY (USA): Wiley*, 2010.

13. SEARLE, N., and WHITE, G., 'Business Models', in *Handbook on the Digital Creative Economy, Edward Elgar Publishing*, pp. 45-56, 2013.

14. BOCKEN, N M P., RANA, P., and SHORT, S W., 'Value mapping for sustainable business thinking', *Journal of Indusrtial and Production Engineering, vol. 32, no. 1*, pp. 67-81, 2015.

15. SPARVIERO, S., 'The Case for a Socially Oriented Business Model Canvas: The Social Enterprise Model Canvas', *Journal of Socieal Entrepreneurship, vol. 10, no. 2*, pp. 232-251, 2019.

16. NIELSEN, C., LUNCH, M., MONTEMARI, M., PAOLONE, F., MASSARO, M., and DUMAY, J., 'Business Models: A research overview', *Routledge*, 2018.

17. MOORE, J F., 'Predators and Prey: A new ecology of competition', *Harvard Business Review*, 1993.

18. PORTER, M E., 'Competitive advantage: Creating and sustaining superior performance', *Free Press & Collier MacMillan*, 1985.

19. ADNER, R., 'Ecosystem as Structure: An actionable construct for strategy', *Journal of Management, vol. 43, no. 1*, pp. 39-58, 2017.

20. JACOBIDES, M G., CENNAMO, C., and GAWER, A., 'Towards a theory of ecosystems', *Strategic Management Journal, vol. 39, no. 8*, pp. 2255-2276, 2018.

21. PHILIPS, M A., and RITALA, P., 'A complex adaptive systems agenda for ecosystem research methodology', *Technological Forecasting and Social Change, vol. 148*, 2019.

22. ZIAJKA-POZNANSKA, E., and MONTEWKA, J., 'Costs and Benefits of Autonomous Shipping - A Literature Review', *Applied Sciences, vol. 11, no. 10*, p. 4553, 2021.

23. GUSENBAUER, M, and HADDAWAY, N R., 'Which academic search systems are suitable for systematic reviews or meta-analyses? Evaluating retrieval qualities of Google Scholar, PubMed, and 26 other resources', *Research Synthesis Methods, vol. 11, no. 2*, pp. 181-217, 2020.

24. EASTMAN, C M., and JANSEN, B J., 'Coverage, relevance, and ranking: The impact of query operators on Web search engine results', *ACM Transactions on Information Systems (TOIS), vol. 21, no. 4*, pp. 383-411, 2003.

25. DOMENIGHINI, C., 'Autonomous inland shipping: will the barge owner be stuck 'between the devil and the deep blue sea'?', *Il Diritto Marittimo, vol. 4, no. 1*, pp. 752-784, 2022.

26. KIM, D., LEE, C., PARK, S., and LIM, S., 'Potential Liability Issues of AI-Based Embedded Software in Maritime Autonomous Surface Ships for Maritime Safety in the Korean Maritime Industry', *Journal of Marine Science and Engineering, vol. 10, no. 4*, 2022.

27. TSVETKOVA, A., HELLSTROM, M., and RINGBOM, H., 'Creating value through product-service-software systems in institutionalized ecosystems – The case of autonomous ships', *Industrial Marketing Management, vol. 99, no. 1*, pp. 16-27, 2021.

28. SANDVIK, H O., SJODIN, D., BREKKE, T., and PARIDA, V., 'Inherent paradoxes in the shift to autonomous solutions provision: a multilevel investigation of the shipping industry', *Service Business, vol. 16, no. 1*, pp. 227-255, 2022.

29. MUNIM, Z H., 'Autonomous ships: A review of innovative applications and future marimtime business models', *Supply Chain Forum: An International Journal, vol. 20, no. 4*, pp. 266-279, 2019.

30. DOMENIGHINI, D., 'Autonomous inland navigation: a literature review and extracontractual liability issues', *Journal of Shipping and Trade, vol. 9, no. 14*, 2024.

31. KATSIVELA, M., 'Legal, Ethical and Industry Issues and Developments regarding Maritime Autonomous Surface Ships', *Ocean Yearbook Online, vol. 37, no. 1*, pp. 506-541, 2023.

32. GHADERI, H., 'Autonomous technologies in short sea shipping: trends, feasibility and implications', *Transport Reviews, vol. 39, no. 1*, pp. 152-173, 2019.

33. KURT, I., and AYMELEK, M., 'Operational and economic advantages of autonomous ships and their perceived impacts on port operations', *Maritime Economics & Logistics, vol. 24, no. 1*, pp. 302-326, 2022.

34. KURT, I., and AYMELEK, M., 'Operational adaptation of ports with maritime autonomous surface ships', *Transport Policy, vol. 145, no. 1*, pp. 1-10, 2024.

35. KRAUSE, S., WURZLER, L., MØRKRID, O E., FJØRTOFT, K., PSARAFTIS, H N., VILANOVA, M R., ZIS, T., COELHO, N F., TATENHOVE, J V., RAAKJÆR, J., KLOCH, K., BILLESØ, M B., and KRISTIANSEN, J N., 'Development of an advanced, efficient and green intermodal system with autonomous inland and short sea shipping – AEGIS', *Journal of Physics: Conference Series, vol. 2311, no. 1*, p. 012031, 2022.

36. NESHEIM, D A., KRISTIANSEN, J N., RAAKJAER, J., and MORKRID, O E., 'The Need for Collaborative Business Models in Autonomous Vessel Ecosystems', in *The 4th International Conference on Maritime Autonomous Surface Ships, Singapore*, 2022.

37. PSARAFTIS, H N., ZIS, T P V., and VILANOVA, M R., 'The logistics of autonomous shipping and the AEGIS project', in *The 8th International Symposium on Ship Operations, Management & Economics (SOME), Athens, Greece*, 2023.

38. RODSETH, O J., PSARAFTIS, H N., KRAUSE, S., RAAKJÆR, J., and COELHO, N F., 'AEGIS: Advanced, efficient and green intermodal systems', *IOP Conference Series: Materials Science and Engineering, vol. 929, no. 1*, p. 012030, 2020.

39. AHN, J., JOUNG, T., KANG, S., and LEE, J., 'Changes in container shipping industry: Autonomous ship, environmental regulation, and reshoring', *Journal of International Maritime Safety, Environmental Affairs, and Shipping, vol. 3, no. 3-4*, pp. 21-27, 2019.

40. MAKKONEN, H., NORDBERG-DAVIES, S., SAARNI, J., and HUIKKOLA, T., 'A contextual account of digital servitization through autonomous solutions: Aligning a digital servitization process and a maritime service ecosystem transformation to autonomous shipping', *Industrial Marketing Management, vol. 102, no. 1*, pp. 546-563, 2022.

41. MASSterly. [Online]. Available: https://www.massterly.com/.

42. BCTN. [Online]. Available: https://bctn.nl/en/our-services/.

43. BCTN, 'Terminal Meerhout', [Online]. Available: https://bctn.nl/terminals/terminal-meerhout/.

44. ANTWERP-BRUGES, P O., 'Partnership between Duisport and Port of Antwerp-Bruges', 17 04 2023. [Online]. Available: https://www.portofantwerpbruges.com/en/news/partnership-between-duisport-and-port-antwerp-bruges.

45. MAGPIE Consortium, 'MAGPIE Project - Work Package presentation', [Online]. Available: https://www.magpie-ports.eu/work-package-presentation/. [Accessed 29 08 2024].

46. Yara, 'MV Yara Birkeland', [Online]. Available: https://www.yara.com/news-and-media/media-library/press-kits/yara-birkeland-press-kit/.

47. Shipping Watch, 'Ikea buys its own containers and charters its own vessels', 2 09 2021. [Online]. Available: https://shippingwatch.com/carriers/Container/article13248779.ece.

48. Business Insider, 'Amazon is making its own shipping containers — and waiting as little as 2 days outside ports', 06 12 2021. [Online]. Available: https://www.businessinsider.com/amazon-making-own-shipping-containers-bypassing-port-delays-2021-12#:~:text=Amazon%20has%20been%20making%20its,span%20up%20to%20two%20months..

8. AUTHORS BIOGRAPHY

Kristoffer Kloch holds the current position as PhD Fellow in an academic-industrial partnership between Aalborg University and DFDS A/S. The Author's previous experience includes working as an officer onboard large commercial ships in international trade, followed by various roles within innovation and partnerships in the maritime industry.

CLASSIFICATION AND DEFINITIONS OF UNCREWED MARITIME SYSTEMS

M I Lesse, Naval Postgraduate School, United States of America
ENS T Milio, United States Navy
D Ponkratov and **J P Martinez,** Royal Institution of Naval Architects, UK

SUMMARY

As uncrewed systems (UxS) permeate the maritime environment, it has become increasingly evident that existing policy lags far behind technological advances. The development of technology, its productive use, and the safety of the maritime environment depend on the existence and implementation of clear laws and policies. This report aims to explore the following:

- Current international definitions of vessels and ships;
- Whether UxS fall within these definitions;
- Current international and FVEY classifications of UxS, or lack thereof; and
- The implications of these classifications (or lack thereof).

Recent seizures of United States (US) UxS by Iran and China deem the classification of these systems as relevant and time sensitive. The findings of this report elucidate the terminology used and precedents set by countries to clarify the legal environment and how UxS will be treated under international law.

NOMENCLATURE

COLREGS	Convention on the International Regulation for Preventing Collisions at Sea
ECS	Extended Continental Shelf
FVEY	Five Eyes Alliance
IMO	Internatinoal Maritime Organization
ITN	Interim Technical Note
MARPOL	International Convention for the Prevention of Pollution from Ships
MASS	Maritime Autonomous Surface Ships
ROV	Remotely Operated Vehicles
UNCLOS	United Nations Convention on the Law of the Sea
U.K.	United Kingdom
U.S.	United States
UxS	Uncrewed Systems

1. INTRODUCTION

As uncrewed systems (UxS) permeate the maritime environment, it becomes increasingly evident that existing policy lags far behind technological advances. Taking effect in 1982, the most commonly accepted regulatory policy on the high seas is the United Nations Convention on the Law of the Sea (UNCLOS) [1]. UNCLOS provisions have an underlying assumption that ships and vessels (terms used interchangeably within the written agreement) must be operated by a human [2]. Because UNCLOS fails to explicitly define either of the terms 'vessels' and 'ships', it is unclear whether or not the convention's provisions apply to uncrewed maritime systems, which could also be interpreted as devices. Lacking an international classification of UxS as ships or vessels, the precedents that individual nations set through their own domestic policies and treatment of UxS become increasingly important in determining the treatment of UxS in the maritime environment [3]. Whether UxS are considered ships or vessels, or if they fall into a separate category, dictates whether UxS are required to follow certain maritime rules and whether they receive specific protections. The development of technology, its productive use, and the safety of the maritime environment depend on the existence and implementation of clear laws and policies. This paper explores the following key elements of the classification of UxS in the maritime environment:

- Current international and Five Eyes Alliance (FVEY) members' definitions of vessels and ships;
- Whether UxS fall within these definitions;
- Current international and FVEY classifications of UxS, or lack thereof; and
- The implications of these classifications (or lack thereof).

In the context of recent seizures of U.S. UxS by Iran and China, the classification of these systems is relevant and time sensitive [4][5]. The findings of this paper elucidate the terminology used and precedents set by FVEY countries to clarify the legal environment and how UxS will be treated under international law. This analysis becomes even more relevant as state boundaries shift around the world. Recently, the U.S. announced the extension of the outer limits of its continental shelf [6]. The U.S. has extended continental shelf (ECS) in seven offshore areas including the Arctic, Atlantic (east coast), Bering Sea, Pacific (west coast), Mariana Islands, and two areas in the Gulf of Mexico. The ECS is approximately one million square kilometers, an area about twice the size of California. The State Department notes that the U.S. "may also have ECS in other areas, and the U.S. ECS Project continues to analyze available data and undertake analysis in a range of areas [7]."

Research relevant to this analysis includes a plethora of legal and policy references, from treaties to military handbooks to ship registration regulations. For this paper, a nation has defined the terms 'ship' or 'vessel' when the term(s) is included in the laws or policies that apply broadly to their coastal waters, territorial waters, or exclusive economic zone, promulgated by the regulatory body in charge of that area (e.g., the U.S. Code of Law, the Canada Shipping Act of 2001, etc.). Specific treaties often have narrow definitions that only make sense in their specific context which don't necessarily offer clarity of legal treatment for UxS [8]. Given the ambiguity among states' treatment of UxS, a broader approach to researching the issue is needed including examining a wide variety of documents and statements to gauge nations' stances on the status of UxS as ships or vessels.

2. DEFINITIONS OF 'VESSEL' AND 'SHIP' AMONG FVEY NATIONS

UNCLOS provisions apply to either 'vessels' or 'ships', though neither of which are defined in the convention itself. However, many nations have delineated between the two terms within their own national legal frameworks. As the International Maritime Organization (IMO) considers rules to fill the gaps under UNCLOS, it is necessary to understand the current global consensus on the definitions of these two terms and whether a common definition may be attained [9]. In the absence of an international agreement on 'ship' and 'vessel'definitions, the precedent set by nations' definition of 'vessel' or 'ship' and whether UxS is included will influence the definition that the IMO eventually uses in its rules. This paper focuses on FVEY countries, as they are incorporating UxS technology into their fleets and exert significant political influence internationally.

2.1 VESSELS

Vessels and ships are referred to interchangeably in UNCLOS and lack formal definitions. It is useful, however, to note that the term 'vessel' has been defined in other internationally accepted maritime legal frameworks. The Convention on the International Regulation for Preventing Collisions at Sea (COLREGs) defines a vessel as "includ[ing] every description of watercraft, including non-displacement craft, wing-in-ground-effect craft and seaplanes, used or capable of being used as a means of transportation on water [10]." The following section discusses FVEY definitions for the term 'vessel.' FVEY member countries consist of the United States, United Kingdom, Australia, Canada and New Zealand.

In the U.S. Code that was legally codified in 1947, 'vessel' is defined broadly in Title 1 as "every description of watercraft or other artificial contrivance used, or capable of being used, as a means of transportation on water [11]." The term is further defined in the U.S. Code Title 47 from 1954 and is equated to the term 'ship,' and uses the exact wording as the first definition except that it excludes the mention of 'aircraft [12].' Interestingly, even in the same code of law, there is ambiguity regarding the distinction between a 'vessel' and a 'ship.'

In the United Kingdom (U.K.), the Merchant Shipping Act of 1894 defines a vessel as "includ[ing] any ship or boat, or any other description of vessel used in navigation," circularly using the term 'vessel' in its own definition [13]. The more recent Merchant Shipping Act of 1995 does not include a definition for the term 'vessel.'

Australia has two main legislative frameworks– one for ships that travel internationally (Regulated Australian Vessels) and another for ships that operate exclusively within the Australian Exclusive Economic Zone (Domestic Commercial Vessels). Regulated Australian Vessels fall under the Navigation Act of 2012, while Domestic Commercial Vessels are regulated by the Marine Safety National Law Act of 2012 [14]. The Navigation Act defines a vessel as "any kind of vessel used in navigation by water, however propelled or moved, and includes the following: (a) a barge, lighter or other floating craft; (b) an air-cushion vehicle, or other similar craft, used wholly or primarily in navigation by water [15]." In the Marine Safety National Law Act, a vessel is defined as "a craft for use, or that is capable of being used, in navigation by water, however propelled or moved, and includes an air-cushion vehicle, a barge, a lighter, a submersible, a ferry in chains and a wing-in-ground effect craft [16]." It is notable that the first Navigation Act defines 'vessel' circularly, using the term in its own definition, while the Marine Safety National Law does not.

The Canada Shipping Act of 2001, the umbrella act under which more specific maritime provisions fall, defines a vessel as "a boat, ship or craft designed, used or capable of being used solely or partly for navigation in, on, through or immediately above water, without regard to method or lack of propulsion, and includes such a vessel that is under construction. It does not include a floating object of a prescribed class [17]." New Zealand's primary marine policy, the Maritime Transport Act of 1994, does not provide a definition for the term 'vessel.'

2.2 SHIPS

The term 'ship' is not defined in UNCLOS but is defined in the International Convention for the Prevention of Pollution from Ships (MARPOL). MARPOL defines ships as "a vessel of any type whatsoever operating in the marine environment….[18]" While this is an example of an international definition in the absence of one in UNCLOS, it is much broader than usual, as pollution regulations (such as MARPOL) aim to include as wide a range of watercraft as possible. Each of the FVEY countries also has a domestic legal definition for the term 'ship.'

In the U.S., 'ship' and 'vessel' are defined interchangeably in the U.S. Code of Law as mentioned in the previous section on vessels. The only difference between the definition of 'vessel' and that of a 'vessel or ship' is that ships do not include aircraft [12].

The U.K.'s Merchant Shipping Act of 1995 defines 'ship' as "every description of vessel used in navigation [20]." Since the Merchant Shipping Act of 1995 does not define a vessel, we can assume that it references back to the 1894 act, which does define the term. Problematically, the 1894 definition of the term 'vessel' is a "ship… used in navigation [13]." As a result, a ship is a type of vessel, while a vessel is also a type of ship under U.K. law.

Neither of Australia's main governing maritime policies, the Navigation Act of 2012 and the Marine Safety National Law Act of 2012, contain a definition for 'ship [16].' The Australian Shipping Registration Act of 1981, however, does provide a definition for 'ship' as "any kind of vessel capable of navigating the high seas…" and includes various examples to clarify items that can be defined as ships [20].

In the Canadian Shipping Act of 2001, there is no provided definition for 'ship.' In the Canada Marine Act, a regulation administered by Transport Canada (the government's commercial regulatory body), it is defined as "every description of vessel, boat or craft designed, used or capable of being used solely or partly for marine navigation, whether self-propelled or not and without regard to the method of propulsion, and includes a sea-plane and a raft or boom of logs or lumber [21]." In New Zealand, under the Maritime Transport Act of 1994, 'ship' is defined as "every description of boat or craft used in navigation, whether or not it has any means of propulsion…" and include examples of items that are to be considered ships [22].

3. COMPLEXITY OF DEFINING UxS

Given that the FVEY countries have defined either 'vessel' and/or 'ship,' the next step in the analysis is whether UxS fall within these definitions and can be categorized as such. Every country, excluding New Zealand and Australia's Marine Safety National Law, uses the term 'vessel' in their definition of a ship, so this analysis will focus on how an UxS can be classified as a 'vessel'. New Zealand's definition for a 'ship' will be used because its law lacks a definition for 'vessel.' Most seem broad and inclusive of UxS at the surface level, but there are various nuances within each definition.

The first issue arises in the existence of multiple definitions of 'vessel,' which use the term to define itself. The circular nature of these definitions makes it difficult to determine whether they are inclusive of UxS. To ascertain whether UxS are vessels, other requirements that are common between all FVEY definitions may help, regardless of whether they use the word 'vessel.' The main specifications are that (1) a vessel is in or on the water in some capacity and (2) they are either capable of or are being used for navigation or transportation.

There is little debate over whether UxS satisfies the first condition of being in or on the water. The second condition, what classifies use for navigation or transportation, is less clear.

For example, in the U.S. Code of Law's definition of 'transportation' provided in Chapter 51, which lays out the conditions of the Transportation of Hazardous Material, states that 'transportations' is "the movement of property and loading, unloading, or storage incidental to the movement." Purely based on the U.S. definition of 'vessel' and the U.S. Code of Law's definition of 'transportation,' a small UxS that looks nothing like a conventional vessel could be defined as such, even if all it is 'transporting' is a small sensor, like the payload of a Wave Glider [23]. However, a more narrow definition of 'transportation' could have excluded the Wave Glider if it implied the carrying of people or goods, or had some implication or requirement of the intentional movement of objects from a start point to an end point (which a sensor does not necessarily do). The way 'navigation' is defined could also exclude various UxS; there are certain propulsion requirements for navigation that may not be met because some UxS move passively through their environment.

It has been up to individual nations to interpret these definitions, determining whether UxS are registered as ships or vessels, and apply their existing policies to the uncrewed systems or create new provisions specifically for UxS. The following section will outline the ways in which FVEY countries have integrated UxS into a policy framework that was written before the existence of such systems.

4. CLASSIFICATIONS OF UxS AMONG FVEY NATIONS

Pending international clarification, states self-determine if UxS are included under existing policy [8]. A study has been conducted by Comite Maritime International in which they asked seventeen countries a series of questions about the status of uncrewed ships [2]. While this study outlined the ways in which different countries interpreted their laws in regard to larger UxS, such as cargo ships, it did not show more broadly how countries are integrating various types of UxS into their policy framework through recent policies, regulations or other actions. This section will outline the provisions promulgated and actions taken by nations to make clear their views on UxS and how such systems fit into an environment of ships and vessels.

4.1 Existing Classifications

The U.S. has explicitly stated in an official publication that UxS are considered ships and are granted the associated maritime rights. In the military context, the 2022 Commander's Handbook on the Law of Naval Operations, UxS are said to have the ability to "operate independently as a ship" and to "exercise any internationally lawful use of the seas [24]." UxS are clearly defined as 'ships,' and therefore fall under the regulatory regime of UNCLOS and other relevant maritime laws. Additionally, the handbook states that UxS "engaged exclusively in government, noncommercial service are sovereign immune craft," solidifying the position that the rights afforded to conventionally crewed ships under UNCLOS also apply to UxS [24]. The U.S. has integrated UxS into an existing policy framework that was created with only crewed vessels in mind at the time the policies were written.

The U.K., while lacking an explicit regulation declaring UxS as vessels, has considered how the modern technology fits into the existing UNCLOS framework. In an International Relations and Defence Committee meeting on UNCLOS and modern uses of the sea, the conclusion reached was that the provisions ought to be interpreted on a "principle of equivalence." This means that if this technology meets the given safety requirements for conventionally crewed ships, they too, should be considered ships and given the associated rights. In 2021 alone, the U.K. registered 23 UxS as vessels [9]. In its Maritime 2050 Strategy, the Department for Transport aimed to be a leader in the "uptake of smart shipping technologies" and in being the "register authority for autonomous and semi-autonomous vehicles," indicating that it views UxS as ships or vessels capable of registration as such [25]. The U.K. is setting a strong precedent by integrating UxS under their policy as vessels.

The Australian Maritime Safety Authority has consistently referred to UxS as 'vessels' when referencing them, and has stated that "currently, these vessels are subject to the same regulatory framework as other vessels, including for survey standards and crewing requirements [26]." The Safety Authority states explicitly that because of the broad definition of 'vessel' in the Navigation Act of 2012 and the Marine Safety National Law Act of 2012, these regulations apply to UxS. Australian authorities have solidly stated that UxS fall under the same regulatory regime as conventionally crewed vessels and have set a precedent that the uncrewed systems are to be treated as vessels.

Transport Canada has created a policy on the Oversight of small Maritime Autonomous Surface Ships (MASS) [27]. Even in the title of the policy, they refer to the UxS as 'ships,' indicating that they will treat them as such in following legal discourse. Additionally, in a 2002 court case regarding an uncrewed submersible device, a Canadian federal court applied the definition of vessel widely, where the physical presence of a crew on board is not necessary to be classified as a vessel [2].

New Zealand issued an Interim Technical Note (ITN) in 2020 to provide guidance for the use of UxS. The ITN refers to the UxS consistently as ships, and provides viable pathways for ship registration, indicating New Zealand's view that UxS are indeed considered to fall under their definition of a 'ship [28].' The document acknowledges that UxS do not meet certain crewing and watchkeeping requirements laid out in other policies and regulations, in which case ship owners are able to apply for exemptions that will allow their UxS to be registered. Interestingly, the ITN excludes "remotely operated vehicles (ROVs) such as those used in subsea operations, which during the mission are physically or wirelessly connected to the relevant support vessel from which they are controlled [28]." The exclusion of this kind of watercraft raises interesting questions and could offer some insight into what 'navigation' and to 'navigate' mean under New Zealand maritime law. Perhaps these vessels are excluded because of the unreliability of their "ability to navigate." If there was any interference with their remote-control communications, this ability would be lost, as opposed to a pre-programmed UxS that does not need constant communication with an operator.

Every FVEY country has either explicitly labeled UxS as 'ships' or 'vessels,' or indicated that the systems receive the same rights under UNCLOS. While it is unclear whether UxS fall under their legal definitions of 'ship' and 'vessel,' these nations have made their interpretation of the law clear by stating so in policies or registering UxS as ships. FVEY nations are setting a precedent for the application of UNCLOS to UxS, giving the uncrewed systems the same rights and responsibilities as conventionally crewed vessels. These include the application of legal application of the right to innocent passage, due regard, and other rights. The application of these maritime rights has become increasingly relevant with UxS seizure incidents like the 2016 USS Bowditch incident [5]. Without a legal regime establishing how UxS should obey certain maritime laws and be treated, countries run the risk of creating an unsafe maritime environment for commercial operation and elevating the potential for military escalation. FVEY countries have effectively established that having humans on a watercraft is not a prerequisite to it being a vessel or ship. While it is encouraging that all FVEY countries are considering UxS and setting a precedent for international law, many questions about their integration into the maritime environment remain. The following table offers a quick reference for these findings in **Table 1**. Note that 'circular' refers to whether the term 'vessel' is used in its own definition.

Table 1. FVEY country considerations of UxS classifications

	Define vessel?	**Define ship?**	**Circular 'vessel' definition?**	**Condition for being a 'vessel' besides being in the water**	**Classify UxS as vessels or ships?**
United States	Y	Y	N	Capable of transportation	Yes, in 2022 Commander's Handbook on the Law of Naval Operations, ship registration
United Kingdom	Y	Y	Y	Used for navigation	Yes, in International and Defence Committee meeting, ship registration
Australia	Y	Y	Y	Used for navigation	Yes, in a formal statement by the Australian Maritime Safety Authority
Canada	Y	Y	N	Capable of navigation	Yes, in Oversight of small Maritime

					Autonomous Surface Ships (MASS) Policy, legal precedent
New Zealand	N	Y	Y	Used for navigation	Yes, in a 2020 Interim Technical Note

5. CONCLUSION AND RESEARCH RECOMMENDATIONS

The research team identified several gaps related to defining UxS and recommends the following research questions to add clarity to international policy around UxS.

How will novel technology be integrated into existing policy? How will new policy be formulated to allow room for technological development?

While this research made clear that UxS can be considered vessels and/or ships despite definitional challenges in existing policy, questions remain about the extent to which this classification will apply. Certain UxS have a clear crewed counterpart, such as an autonomous cargo ship or frigate versus a conventional cargo ship. Other UxS, however, will be novel technology in their appearance, use, and capabilities. It will be significantly more difficult to integrate UxS into existing policy when the systems do not have a conventionally crewed counterpart to which they can be compared. What this analysis has shown is that in FVEY countries, being labeled as a vessel does not necessitate the level of crew present. What it has not shown is whether all UxS, present and future, will be considered UxS by these nations. The extreme variability in these systems, much of it still unknown, will continue to pose legal and policy challenges.

What does it mean to 'navigate' or to 'transport?'

Technology is fundamentally changing the ways in which navigation happens and what it means to do so. For example, the Wave Glider can navigate without a means of mechanical propulsion as the system simply harnesses the wave energy in the surrounding water to move in its intended direction [23]. Additionally, there is the question of whether the small sensors that systems, like those on Wave Gliders, carries constitute 'transportation.' Seemingly simple terms leave ambiguity in how intentional navigation needs to be (to what extent is a device drifting with the waves 'navigating'); by what means the navigation needs to occur (are there requirements for propulsion); and what degree of autonomy the navigation needs to occur at (does a remote control equate to an artificially intelligent craft). While a precedent has been set that UxS do indeed fall under the requirements of 'navigating' and 'transporting' enough to be considered vessels, New Zealand's decision to exclude UxS that are remote controlled shows that there will be much more nuance involved as modern technology emerges. Clear definitions of these terms will be necessary, as well as further classifications of UxS by degree of autonomy, networking capability, intended operational use, and a variety of other characteristics.

How do non FVEY countries, especially U.S. adversaries, define UxS?

The ambiguity surrounding UxS can be taken advantage of by other nations to commit legally questionable actions. Without the same protections that are given to crewed vessels and ships in UNCLOS, UxS are subject to unfair treatment by other hostile nations. The U.S. has encountered at least two documented incidents where two countries seize its UxS (China and Iran). In each case, both countries claimed that their actions were legal and justified with accordance to their domestic and international maritime laws. In future research, it would be interesting to uncover more about how these nations define UxS. If they do not consider UxS ships or vessels under their laws, UNCLOS would not apply. If they do, they were simply acting provocatively by seizing the UxS and taking actions they knew were legally ambiguous. For instance, there has since been a revision to New Zealands ITN issued in December of 2023 that should be further explored to see if definitions have been updated [29]. While this report has shown that FVEY countries are setting a precedent of treating UxS as vessels legally, other nations may not be doing the same.

6. ACKNOWLEDGEMENTS

We would like to thank the Consortium for Robotics and Unmanned Systems Education Research (CRUSER) group within the Naval Postgraduate School for continuing to fund our research into important legal and policy concepts of UxS. The opinions expressed within this paper are solely those of the authors.

7. REFERENCES

1. UN General Assembly, 'Convention on the Law of the Sea, Article 29', *UNCLOS*, 1982.

2. Comite Maritime International, 'Summary of Responses to the CMI Questionnaire', *CMI International Working Group Position Paper on Unmanned Ships and the International Regulatory Framework*, 2018.

3. SCHMITT, M., and GODDARD, D R., 'International law and the military use of unmanned maritime systems', *International Review of the Red Cross*, 2016.

4. GAMBRELL, J., 'Iran briefly seizes 2 US sea drones in Red Sea amid tensions', *Associated Press*, 2022.

5. FRAVEL, M T., 'The Implications of China's Seizure of a U.S. Navy Drone', *The National Bureau of Asian Research (NBR),* 2016.

6. Office of the Spokesperson, 'Announcement of U.S. Extended Continental Shelf Outer Limits', *U.S. Department of State*, 2023.

7. U.S. Department of State, 'U.S. Extended Continental Shelf', *U.S. Department of State*, 2023.

8. MCKENZIE, S., 'When is a Ship a Ship? Use by State Armed Forces of Un-crewed Maritime Vehicles and the United Nations Convention on the Law of the Sea', *Melbourne Journal of International Law*, 2020.

9. International Relations and Defence Committee, 'Corrected oral evidence: UNCLOS: Fit for purpose in the 21st century?', *UK Parliament*, 2021.

10. International Maritime Organization, 'Convention on the International Regulations for Preventing Collisions at Sea, Rule 3', *OpAnalytics*, 1972.

11. United States Code, 'Rules of Construction, 1 U.S.C.§ 3', *Cornell Law School*, 1947.

12. United States Code, 'Definitions, 47 U.S.C. § 153', *Cornell Law School*, 1954.

13. UK Parliament, 'Merchant Shipping Act of 1894', *UK Parliament*, 1894.

14. Australian Maritime Safety Authority, 'Advisory note – The scope of the National System: Regulated Australian vessels and domestic commercial vessels', *Australian Government*, 2015.

15. Department of Infrastructure, Transport, Regional Development, Communications and the Arts, 'Navigation Act 2012', *Australian Government*, 2012.

16. Department of Infrastructure, Transport, Regional Development, Communications and the Arts, 'Marine Safety (Domestic Commercial Vessel) National Law Act 2012', Australian Government, 2012.

17. Senate and House of Commons of Canada, 'Canada Shipping Act', *Government of Canada*, 2001.

18. International Maritime Organization, 'International Convention for the Prevention of Pollution from Ships, Article 2 Section 4', *International Maritime Organization*, 1973.

19. UK Parliament, 'Merchant Shipping Act of 1995', *UK Parliament*, 1995.

20. Department of Infrastructure, Transport, Regional Development, Communications and the Arts, 'Shipping Registration Act 1981, *Australian Government*, 1981.

21. Senate and House of Commons of Canada, 'Canada Marine Act', *Government of Canada*, 1998.

22. Parliamentary Counsel Office, 'Maritime Transport Act 1994', *New Zealand Legislation*, 1994.

23. Liquid Robotics, Inc., 'The Wave Glider | How It Works', *Liquid Robotics, Inc.*, 2022.

24. Office of the Chief of Naval Operations, 'The Commander's Handbook on the Law of Naval Operations', *U.S. Department of Navy*, 2022.

25. Department for Transport, 'Maritime 2050 - Navigating the Future', *U.K. Department for Transport*, 2019.

26. Australian Maritime Safety Authority, 'Autonomous vessels in Australia', *Australian Government*, 2022.

27. Transport Canada, 'Tier I- Policy- Oversight of small Maritime Autonomous Surface Ships (MASS)', *Government of Canada*, 2022.

28. Maritime New Zealand, 'Autonomous Ship Operation in New Zealand', *Government of New Zealand*, 2020.

29. Maritime New Zealand, 'Rev. 2, Autonomous Ship Operation in New Zealand', *Government of New Zealand*, 2023.

8. AUTHORS BIOGRAPHY

Marina Lesse is a faculty research associate in the EAG at the Naval Postgraduate School. Her background is in marine and environmental science and robotics and unmanned systems. Ms. Lesse's expertise includes the utilization of uncrewed underwater and surface vehicles for ocean research. Additionally, her expertise includes environmental education, habitat restoration, watershed conservation, and marine biology. She has a BS in marine science from California State University–Monterey Bay and is pursuing an MA in Strategic Studies in the NPS National Security Affairs department. Lesse was an NPS–Stanford Climate Security Fellow in the inaugural 2023-2024 class and is now a senior fellow.

Theresa Milio graduated with honors from the United States Naval Academy in 2023 with a BA in Political Science. While interning at the Naval Postgraduate School in 2023, she researched the definitions of uncrewed systems and the implications for international law. She is now serving in the U.S. Navy as a submarine officer.

ENVIRONMENTAL POTENTIAL OF REMOTE CONTROL TECHNOLOGIES: USE CASE REMOTE PILOTAGE

M Yildiz, **N Hagemeister**, **R Grundmann** and **H C Burmeister**, Fraunhofer CML, DE, Germany

1. ABSTRACT

Remote-control technologies for maritime applications are emerging. Despite ongoing international efforts on MASS regulation, regulatory uncertainties remain a hurdle. Besides remote control, remote support in terms of shore-based pilotage can provide a quicker way to exploit the benefits of these technologies' advancements. If virtual boarding of the vessel becomes possible – at least in certain cases – the need for pilot transfer by vessel or helicopter becomes obsolete. This paper briefly introduces relevant MASS technologies for enabling remote pilotage services, before assessing its environmental impact reduction potential by deriving the annual profile of pilot boat missions within Europe based on one year of AIS-data. Hereby, the speed and size profiles are used for fuel oil consumption estimates to derive a realistic estimate for the potential savings per coastal state. The paper concludes by discussing operational boundary conditions for remote pilotage services and comparing the findings with an earlier study on pilotage activities.

2. CURRENT PILOTAGE PROCESS

Pilotage in the maritime sector represents an indispensable component of navigation, presenting numerous challenges. A pilot must not only possess extensive knowledge of local waters but also be capable of making precise decisions under extreme conditions. One of the greatest challenges is navigating narrow and heavily trafficked waterways. Here, the pilot must safely guide unfamiliar vessels through tight passages, under bridges, around port facilities, and finally to the berth. These tasks are often performed under limited visibility and changing weather conditions. Detailed knowledge of depth conditions, currents, and potential obstacles is essential.

Before entering the harbor, the captain of the piloted vessel must request a pilot. The pilot, being familiar with the local traffic, hydrographic and meteorological conditions, prepares by analyzing traffic, weather, tides, and currents. The boarding of a vessel by a pilot can be done in various ways, depending on prevailing conditions, the type of vessel, weather conditions, and safety regulations. The most common method of pilot transfer is by pilot boat, where the pilot is brought to the ship using a dedicated boat. The pilot boat approaches the ship, and the pilot boards via an appropriate medium. The most common method here is the pilot ladder, where the pilot climbs aboard using a ladder lowered along the side of the ship. This method is often used as it is flexible and adaptable to various ship types and heights. In some cases, a gangway, often combined with a pilot ladder, can be used, particularly with larger ships or when the ship's side is relatively low. Another method is the Pilot Access Door or Pilot Door, a door in the ship's hull. The pilot can directly board the ship from the pilot boat through this door without using a ladder. This method is often used on modern large ships where ladder access would be risky or impractical. In certain regions, especially under difficult weather conditions or at sea, the pilot can also be brought to the ship via helicopter. This is common in areas with heavy seas or with large vessels. The helicopter hovers over the deck of the ship, and the pilot is lowered or picked up by winch.

Upon boarding, an initial information exchange occurs between the pilot and the bridge crew, after which the pilot takes over the nautical advisory role while the captain retains overall command. In the harbor, the pilot navigates the vessel through narrow and busy waterways, constantly communicating with the Vessel Traffic Service (VTS) or shore-based infrastructure officials. The pilot continuously adjusts speed and course to avoid collisions and coordinates tug assistance for precise docking maneuvers. The pilot positions the vessel at the berth and assists the bridge team in coordinating the crew to deploy mooring lines. After secure mooring, the pilot hands responsibility back to the captain and port authorities, disembarks, documents the operation, and, if necessary, writes reports on any special incidents to be forwarded to the relevant authorities. Taking over the pilot, by any means, not only poses an increased risk to human life - despite sophisticated safety measures, the pilot still depends on the quality of work and preparation of others - but also leads to increased emissions from the pilot's means of transportation.

In summary, successful pilotage does thus require two conditions:

1) The situational awareness of the pilot of the vessels and its situation, and
2) The trust of the master or officer of the watch in the pilot's recommendations

Remote pilotage technologies have the potential to make this process more environmentally friendly and at the same time safer for the pilots, but as it *Remote Pilotage is not about making pilotage obsolete, [but] rather [about] providing pilots with modern virtual means to execute their tasks to assist the onboard crew in a safer way* [1] both conditions must also be met by remote pilotage applications. The next chapter will shed light on the availability of technologies to ensure this from the domain of maritime systems before the full environmental potential of remote pilotage technologies for Europe is estimated based on an AIS analysis in the following chapters.

3. AUTONOMOUS MARITIME SYSTEMS FOR REMOTE PILOTAGE

Technologies enabling Autonomous Maritime Systems are on the rise and involve a variety of industrial actors and products meanwhile. Business-wise, enabling *Remote Operations* is one of the main motivations among the identified use cases for this technology [2]. Technology developments for maritime autonomy are often characterized along the four main categories [3]:

- Lookout systems for automated perceived situational awareness,
- Autonomous navigation systems for nautical decision-making,
- Shore-based remote operation centers for human supervision, and
- Technical monitoring systems for engine operations.

While the last section is not really relevant in the case of pilotage, innovations in the other areas do have relevance for enabling remote pilotage. Lookout systems aim at gathering the situation around the vessel by an integrated sensor set-up to enable safe nautical decision-making. While there is still uncertainty how this situation shall be completely described [4], more than 20 products covering this aspect are already available in the market, which can also contribute to gathering information for shore-based pilots [5]. With regards to nautical decision-making, a variety of algorithms developments for e.g. COLREGs-compliant maneuvering [6] or anomaly detection [7] is existing. Despite not being in actively doing decision-making in available products, such methods do possess a potential to assist as decision-recommendation applications in situations as remote pilotage, where a lack of onboard situational awareness need to be covered by alternative means. Finally, the remote operation center developments typically serve as distant human-interaction systems to enable safety and control via a collaborative ship-shore system [8], with mixed reality being on emerging technology in that field [9]. These developments are specifically interesting, as the area of remote operations centers do mainly cover the challenge of ensuring distant situational awareness of shore-based personnel, which is the main hurdle to be overcome for remote pilotage.

Thus, it is also no surprise to see the first prototype testing of remote-based pilotage systems in action, which do use elements of the above-mentioned development. Recent projects in that area are e.g. the Finnish Sea4Value project, which uses data analysis, artificial intelligence (AI) and digital twins to optimize navigation, port operations and pilotage services. Real-time data is used to monitor weather, currents and ship traffic more efficiently, resulting in time and cost savings. [1]. A further approach has been tested by the RePo MAN project, which envisioned overcoming the situational awareness challenge by application of mixed reality to digitize pilotage services and make them more immersive [10].

Specifically for the area of situational awareness, it can thus be concluded that a variety of technology building blocks are already existing or under active development within the autonomous maritime systems domain. A technology gap does primarily remain in the field of mutual decision-making and trust in remote set-ups. Consequently, with technical feasibility becoming less of a challenge for the application of autonomous maritime technologies, its gained technological maturity can also benefit remote pilotage applications on short notice to realize its worker safety and environmental benefits.

4. METHODOLOGY FOR ESTIMATING FUEL CONSUMPTION BASED ON AIS DATA

This study creates insight into the current environmental impact of the ship-based maritime pilotage based on historical AIS data. Through an extensive spatial analysis based on AIS data, it quantifies the potential benefits of remote pilotage in terms of reduced fuel consumption and its subsequent environmental implications. This section outlines the methodology, beginning with the preprocessing of AIS data—incorporating both static and dynamic information—and progressing to the data aggregation and analysis techniques employed in this study.

4.1 DATA DESCRIPTION AND PRE-PROCESSING

Historical AIS data used in this study was acquired from ORBCOMM, spanning the year 2023 and comprising a combination of terrestrial (shore-based stations) and satellite AIS data from Class A and Class B vessels. We specifically chose 2023 to minimize the potential effects of COVID-19 restrictions on our results. To ensure valid comparisons with the analysis presented by [1], we employed the same methodology for preprocessing and aggregating AIS data.

As [1] doesn't explicitly explain the AIS data aggregation and preprocessing methodology, in this work, we aim to facilitate result reproduction by presenting a thorough data preprocessing methodology while also referring to our older study. Our data analysis pipeline consists of three stages:

- **Data Preprocessing**: Preliminary filtering of global AIS data to create a smaller subset based on MIDs and ship type. Removal of erroneous AIS messages considering AIS message standards
- **Data Aggregation**: Further processing of AIS trajectories to aggregation information by country and port.
- **Data Analysis**: Estimation of fuel consumption based on vessel sizes and speeds, following the methodology outlined by [1, 12].

The data preprocessing stage can be detailed as follows:

- Filtering AIS data based on the ship type message broadcast by each vessel. We used ship type "50" to identify pilot vessels, which excludes pilotage conducted by tugs or other vessel types from this study.
- Filtering vessels by Maritime Identification Digits (MIDs) ranging from 201-279 which is reserved for Europe.
- Further refining the MMSI list by removing vessels that reported multiple ship types throughout the year, including but not exclusively type 50. This step aimed to eliminate vessels that may have erroneously broadcast ship type 50 due to unknown reasons.
- Pilotage vessels of length greater than 33 meters have been excluded from the analysis to remove offshore pilotage stations.

Up to this stage, we have mostly used static information of the vessels to filter them out before the final analysis, in the data aggregation part, however, we performed further processing based on the dynamic data (latitude, longitude, SOG, COG, heading) based on features such as spatiality and expected regular temporality. The data aggregation process can be summarized as follows:

- **Land Area Filtering**: AIS points intersecting with land areas were removed to ensure the exclusion of erroneous AIS messages inside land. For defining an AIS message inside a land, we used the spatial boundaries given in dataset [16].
- **Temporal Continuity Check**: Data points were filtered to maintain temporal continuity, retaining only records with time differences between 0 and 600 seconds (10 minutes) between consecutive AIS messages from the same vessel. 600 seconds is chosen according to the standards of AIS data update intervals. [11] Even though the slowest update rate is 3 minutes, 10 minutes is chosen after careful inspection of the data.
- **Distance Calculation**: The distance travelled between consecutive AIS messages from the same vessel was calculated using the haversine distance. AIS messages that are found to have more than 6km distance are filtered out to ensure erroneous AIS reportings are excluded. 6 km is chosen based on the assumption of a pilotage vessel going with constant 20knots, which can be considered as a high speed (see Table 2), for 600 seconds which is the temporal threshold used.
- **Trajectory Chunking**: After the removal of AIS messages that have more than 600 seconds to the previous AIS message, new gaps are created between t_0 and t_2 if we assume t_1 is the trajectory point at the time where a 600-second time difference is observed (t_0, t_1, t_2 shows chronologically three consecutive AIS messages from the same vessel). To ensure that, these gaps are omitted while aggregating the data for fuel consumption analysis which uses the time difference between consecutive AIS messages, we have created chunks of the data where each chunk consists of timely sorted (ascending) AIS messages from the same vessels that are ensured have time difference between consecutive points less than 600 seconds.
- **Pilotage Mission Extraction**: A leg extractor algorithm was applied to identify potential missions

or significant maneuvers, focusing on instances where the heading changed by more than 90 degrees while the vessel was moving (SOG > 1knot). Our leg extraction algorithm is an unsupervised learning-based trajectory classifier, that learns the statistical properties of noise in the transmitted heading values to classify leg intervals (list of consecutive trajectory points) that vessels are performing a maneuver or going straight. Due to the unsupervised learning process, we have used our algorithm on complete trajectories rather than the chunks we have defined.

- **Area Classification**: Area classification in this context refers to assigning pilotage vessels to the specific country they operate. To ensure each vessel is assigned to the correct country, we have used a two-step approach. The first step was to investigate a trajectory's closest point to a known port. A 50 km threshold is used to determine a pilot vessel's origin port, and 50 km is chosen arbitrarily to create a buffer zone around the port. Even though, if the pilot operation happens at a faraway distance from the port due to the distance between ports and pilotage zones, checking the closest point of a pilot vessel to a port allowed us to use berthing times of the pilot vessels for port assignment. Theoretically berthing should happen very close to a port, however since we have not created port locations list ourselves, we assumed a 50 km buffer zone would account for the error between the port location and berthing zones. If a candidate port has not been found via this method, we have used the first three digits of the MMSI to assign a country for the analyzed trajectory. At this point we would like the emphasize that, even though most pilot vessels are expected to berth in a day at least once which would be enough for assignment by port distance, however since we are analyzing a year's worth of Europe-wide AIS data, due to effects such as removal of erroneous SOG values, these berthing points may be lost.
- **Fuel Consumption Estimation**: Fuel consumption was estimated for each data point based on the vessel's length and speed using the methodology from [1, 12]. This methodology which is explained in detail in section 4.2, assumes constant speed between two consecutive AIS messages from the same vessel, therefore we emphasize in certain cases such as maneuvers, this method can overestimate the fuel consumption assuming, most of the maneuver action may not be presented in the data. Table 1 supports this hypothesis, as we can see average time resolutions for many countries are much higher than the AIS update rates during maneuver.
- **Aggregation**: The processed data was aggregated for their corresponding country, compiling statistics on time resolution, total distance traveled, number of position reports, number of pilot missions, speed distributions, and fuel consumption.

This rigorous preprocessing pipeline ensures that the AIS data used in the analysis is clean, relevant, and accurately represents the operational characteristics of pilot vessels in various maritime regions. Our methodology differs from [1] primarily in the "Area Assignment" process. While our previous study utilized a combination of Maritime Identification Digits (MID) and bounding box based area assignment, this study employs a mixture of MID and port locations, as the area of study is significantly larger.

4.2 FUEL OIL CONSUMPTION ESTIMATION METHODOLOGY

Estimation of the fuel oil consumption (FOC) followed the methodology laid out by [1, 12]. A list of pilot vessel models was compiled, including their principal dimensions, displacement, and maximum velocity. The search focused on hard-chine designs optimized for high Froude numbers, which appear to be the most common pilot vessel shape in the geographical area of interest. The vessels in the list ranged from approximately 13m to 24m in length, with an average length of 17.28m. Displacements varied between 16t and 55t, with an average displacement of 28.25t. For some designs, displacement information was unavailable. Maximum Froude numbers and slenderness ratios were calculated based on waterline lengths, which were estimated at 90% of length overall when unavailable. Maximum Froude numbers ranged from 0.84 to 1.23, with an average of 1.03. Slenderness ratios were found to be between 4.63 and 5.94, with a mean value of 5.24.

Since displacement figures are not part of AIS messages, a slenderness ratio of 5.0 was selected as representative of the fleet. Minimum bare hull resistance values for Froude numbers between 0.4 and 1.0 from were normalized by division through dynamic pressure. The normalized resistance values (c_T) were scaled from the notional 500t size used in the publication to the size of each individual vessel by assuming geometric similarity and multiplication with the scale factor squared. The reference length of the 500t reference vessel was determined to be 43.3m based on the chosen slenderness ratio of 5.0. The individual scale factor of each vessel was determined by dividing the length (L) from the AIS message by the reference length of 43.3m. Assuming Froude and Reynolds similarity, the normalized and scaled resistance was determined by interpolation up to Froude numbers of 1.0 beyond which extrapolation was applied. The resistance values were then multiplied by dynamic pressure (ρ/2 * v^2) to derive total resistance. This approach neglects the scaling of the frictional resistance component, which is overestimated by up to about 25%. However, this was deemed acceptable as the goal was to create a rough estimate, and the frictional resistance only represents part of the total resistance. The total resistance was then multiplied by the velocity (v) and divided by an estimated efficiency of 0.65 to determine the required engine power. At this stage a cross check was performed through comparison of the derived power obtained for maximum speed and installed power of some of the pilot vessel models. Finally, the fuel oil consumption was determined by multiplying the required engine power with a Specific Fuel Oil Consumption (SFOC) of 220g/kWh.

$$FOC(v, F_n) = c_T(F_n) * \left(\frac{L}{43.3m}\right)^2 * \frac{\rho}{2} * v^3 * 0.65 * SFOC$$

This methodology provides a comprehensive approach to estimating fuel oil consumption for pilot vessels, considering various vessel characteristics and hydrodynamic principles.

4.3 PILOT VESSEL MISSION EXTRACTION

To estimate the number of pilot missions in our data, we have used a novel method for identifying pilot vessel missions from AIS data, leveraging the characteristic maneuvers performed by these vessels during their operations. The core assumption underlying this methodology is that a pilot vessel typically executes a significant change in heading after completing a mission, such as when returning to port after a pilot transfer has taken place.

The mission extraction process utilizes a specialized leg extractor algorithm which extracts manoeuvre and straight legs from a given AIS trajectory. To distinguish actual missions from routine movements, the following criteria were applied:

- A heading change greater than 90 degrees between the start and end of the maneuver. While a 180-degree turn might be more typical for a completed mission, the 90-degree threshold was chosen as a conservative measure to ensure high recall of potential missions.
- A speed over ground (SOG) of at least 1 knot at both the start and end of the maneuver, indicating active vessel movement rather than drifting or mooring operations.

Choice of SOG and heading change thresholds has been made via experimentation and expert opinions on the subject. For each maneuver meeting these criteria, a mission counter was incremented. This approach provides an upper bound estimate of pilot vessel missions, as it may include some non-mission maneuvers but is unlikely to miss genuine missions. This methodology enables a robust estimation of pilot vessel mission frequency and characteristics based solely on AIS data, without requiring additional information about specific pilot operations. On a side note, our leg extraction algorithm prioritizes unsupervised learning on "heading" values in the AIS messages to ensure noise introduced by external effects such as wind or waves are learned, however in the case that "heading" values are transmitted as None, we have instead used "course over ground (COG)" values recorded in the AIS messages.

4.4 SPEED ANALYSIS

The speed analysis was performed on the AIS data to understand the operational profile of pilot vessels in different areas. The analysis focused on the distribution of time spent at various speed ranges. We assumed that speeds between 0 and 30 knots were relevant for pilot vessel operations, with speeds above 30 knots being grouped into the highest speed bin. The speed range was divided into 30 bins of 1 knot each, with the last bin including all speeds of 30 knots and above. For each area, we calculated the percentage of time spent in each speed bin by summing the time differences between consecutive AIS reports of the same vessel falling within each bin. This approach accounts for varying time intervals between AIS reports, providing a more accurate representation of time spent at different speeds compared to a simple count of speed observations. However, it must be noted that this approach assumes a constant speed between consecutive AIS messages of the same vessel. The analysis was performed for individual areas and aggregated across all areas to provide a comprehensive view of pilot vessel speed profiles.

5. RESULTS AND DISCUSSION

In the initial search for ship type in global AIS data from 2023, we have identified 5652 vessels that reported at least once ship type "50" which is the identifier for pilot vessels. Following this initial number, removals have been made as described as the data preprocessing step in the methodology. Resulting alterations to the dataset are presented as:

- Out of these 5652 vessels, 888 of them found to transmit another ship type throughout the year and removed.
- 1085 Pilot vessels with length greater than 33m or length equals to 0 meter are removed.
- 2343 vessels with MMSI's starting with a different number than "2" are removed to get Europe MIDs.

Finally, the remaining 1336 vessels are thoroughly analyzed according to the steps specified in the data aggregation and data analysis parts of the methodology section. The analysis of AIS data for pilot vessels across various European countries reveals significant insights into their operational patterns, fuel consumption, and potential environmental impact.

5.1 AIS DATA STATISTICS

Table 1 presents a comprehensive overview of AIS data and pilot vessel statistics by country, showcasing the average time resolution, total number of AIS reports, unique vessels, total missions, and missions per vessel. Since MIDs are considered for country assignment, tables present entries such as Faroe Islands and Madeira which can be discussed regarding a strict country definition, however still included due to the fact that they have their own MIDs.

Table 1. Comprehensive AIS Data and Pilot Vessel Statistics by Country (Ordered by Total Missions)

Country	Average Time Resolution (s)	Total Number of AIS reports	Unique Vessels	Total Missions	Missions per Vessel
United Kingdom	136.61	2,415,793	139	5,893	42.40
Germany	120.60	2,559,953	43	5,129	119.28
Denmark	136.64	1,569,663	47	4,510	95.96
Netherlands	126.55	1,805,922	37	4,084	110.38
Spain	160.96	2,138,178	98	4,081	41.64
Sweden	170.87	1,183,774	72	3,579	49.71
Norway	104.21	1,197,487	38	3,361	88.45
Finland	157.62	732,454	60	2,762	46.03
France	173.73	979,447	95	2,825	29.74
Russian Federation	207.79	931,741	73	2,548	34.90
Turkey	261.39	715,566	82	1,835	22.38
Belgium	119.46	567,726	10	1,025	102.50
Romania	127.82	328,975	17	817	48.06
Ukraine	229.79	227,002	25	797	31.88
Estonia	207.63	218,426	18	649	36.06
Poland	186.18	266,255	12	637	53.08
Ireland	147.76	213,284	17	604	35.53
Morocco	160.15	236,339	15	597	39.80
Lithuania	145.55	203,260	4	499	124.75
Malta	210.34	161,155	7	503	71.86
Portugal	191.20	121,251	23	321	13.96
Latvia	171.07	106,455	7	330	47.14
Greece	193.62	112,658	9	298	33.11
Croatia	183.38	95,858	18	279	15.50
Gibraltar	185.41	82,635	2	173	86.50
Iceland	154.08	24,229	7	68	9.71
Bulgaria	162.90	44,644	6	79	13.17
Cyprus	124.27	32,608	6	76	12.67

Slovenia	219.39	26,379	2	56	28.00
Italy	208.75	13,719	6	32	5.33
Georgia	242.92	6,727	3	13	4.33
Faroe Islands	221.55	5,255	3	10	3.33
Albania	154.14	3,422	2	8	4.00
Madeira	176.35	1,435	1	3	3.00
Montenegro	157.68	1,663	3	2	0.67

Germany, the United Kingdom, and Spain emerge as the countries with the highest number of AIS reports in our dataset, which alone however cannot indicate intensive pilotage activities considering the fact actual sources of data or filtering activities made by the provider are unknown. Germany, for instance, recorded 2,559,953 AIS reports from 43 unique vessels, conducting a total of 5,129 missions. This translates to an average of 119.28 missions per vessel, suggesting high utilization of pilot boats in German waters. The United Kingdom shows even higher total missions (5,893) but with a lower missions per vessel ratio (42.40) due to its larger fleet of 139 unique vessels. Spain, while third in total AIS reports (2,138,178), falls between Germany and the UK in terms of fleet size with 98 unique vessels, conducting 4,081 missions at 41.64 missions per vessel. On the other hand Netherlands, which is known to contain some of the busiest ports in Europe, has a lower total number of AIS reports yet the total number of missions still reflects the density of pilotage activities. Germany, Denmark, Belgium, Lithuania and Norway show a higher number of "Missions per Vessel" which may highlight them as candidates for remote pilotage.

Montenegro also presents an interesting case with a “Mission per Vessel” value smaller than 1. Upon further investigation of the MMSIs we have identified for Montenegro, we have observed that one of the pilot vessels was indeed not giving any position update for more than a year, however since she has at least once transmitted its static information, our MID and ship type based filtering included it. Considering the fact that, other countries may also have pilot vessels that are not operating (not performing pilot missions) but still existing in the list, actual “missions per vessel” for each country may be higher than what we have presented here.

5.2 SPEED DISTRIBUTION ANALYSIS

The speed distribution analysis presented in Table 2 offers valuable insights into the operational characteristics of pilot vessels across Europe and is heavily linked to the fuel consumption estimate. This data reveals several key patterns in pilot vessel operations:

- **Dominant Low-Speed Operations:** A significant portion of time (23.82%) is spent at very low speeds between 0 and 1 knots. This likely represents stationary periods, such as when vessels are waiting for assignments, during pilot transfers, or when they are docked.
- **Low-Speed Range:** Cumulatively, 10.41% of the time is spent at speeds between 1 and 3 knots. This includes the stationary periods mentioned above and slow maneuvering, which could indicate operations in port areas or during approach to ships.
- **Mid-Range Speeds:** There’s a relatively even distribution of time spent in the 3-10 knot range, with each 1-knot interval accounting for approximately 5-7% of the total time. This range likely represents a mix of transit operations and active pilotage duties.
- **Higher Speeds:** Operations at speeds above 10 knots are less common but still significant. About 18.92% of the time is spent at speeds between 10 and 20 knots. These higher speeds might be necessary for rapid responses to assignments or for covering longer distances efficiently.
- **High-Speed Operations:** Speeds above 20 knots are relatively rare, accounting for only 4.67% of operational time. This suggests that very high-speed operations are not typical for pilot vessels, possibly due to fuel efficiency considerations or operational restrictions.

Table 2. Speed Distribution Analysis for All Areas

Speed Range (kn)	Percentage of Time	Speed Range (kn)	Percentage of Time	Speed Range (kn)	Percentage of Time
0.0 – 1.0	**23.82%**	10.0 – 11.0	3.45%	20.0 – 21.0	1.08%
1.0 – 2.0	5.73%	11.0 – 12.0	2.25%	21.0 – 22.0	0.90%
2.0 – 3.0	4.68%	12.0 – 13.0	1.63%	22.0 – 23.0	0.78%
3.0 – 4.0	6.65%	13.0 – 14.0	1.42%	23.0 – 24.0	0.60%
4.0 – 5.0	6.51%	14.0 – 15.0	1.38%	24.0 – 25.0	0.45%
5.0 – 6.0	6.57%	15.0 – 16.0	1.44%	25.0 – 26.0	0.34%

Speed Range (kn)	Percentage of Time
6.0 – 7.0	6.96%
7.0 – 8.0	6.56%
8.0 – 9.0	5.57%
9.0 – 10.0	5.02%

Speed Range (kn)	Percentage of Time
16.0 – 17.0	1.36%
17.0 – 18.0	1.45%
18.0 – 19.0	1.62%
19.0 – 20.0	1.28%

Speed Range (kn)	Percentage of Time
26.0 – 27.0	0.23%
27.0 – 28.0	0.16%
28.0 – 29.0	0.08%
29.0 – 30.0	0.05%

This speed distribution might have several implications:

1. **Fuel Consumption:** The significant time spent at very low speeds (below 1 knot) suggests that pilot vessels spend a large portion of their operational time in a highly fuel-efficient state. However, the distribution of speeds above 10 knots, while less frequent, likely contributes disproportionately to overall fuel consumption.
2. **Operational Diversity:** The wide range of speeds used indicates the diverse nature of pilotage operations, from long stationary periods to high-speed transits. This diversity underscores the complex nature of pilot vessel operations and the need for versatile vessels and crews.
3. **Safety Considerations:** The concentration of operations at very low speeds likely reflects the careful maneuvering required during pilot transfers and operations near other vessels or in confined waters.
4. **Fuel Efficiency Opportunities:** While a large portion of time is spent at fuel-efficient low speeds, there may be opportunities for further fuel efficiency improvements in the higher speed ranges (above 10 knots) where significant time is spent and fuel consumption is higher.

5.3 FUEL CONSUMPTION ESTIMATION OF EUROPEAN PILOTAGE SERVICES

Table 3 together with Figure 1 and Figure 2 presents a comprehensive overview of the total fuel consumption and distance covered by pilot vessels for each country in our study. This data offers crucial insights into the environmental impact and operational patterns of pilotage activities across different European countries.

The Netherlands stands out as the country with the highest fuel consumption at 4,193.58 tons, despite not having the greatest distance covered. United Kingdom is found to cover the greatest distance at 827,211.39 nautical miles, with the second-highest fuel consumption of 3339.01 tons. This suggests that Dutch pilot vessels generally operate at higher speeds compared to other countries, leading to increased fuel usage relative to distance covered.

The United Kingdom presents an interesting case, covering a considerable distance (827,211.39 nautical miles, the highest in the dataset) with a fuel consumption of 3,339.01 tons. Denmark shows a moderate fuel consumption (1,887.63 tons) relative to its distance covered (516,975.38 nautical miles), suggesting a balance between operational needs and fuel efficiency.

At the other end of the spectrum, countries like Montenegro and Albania show minimal fuel consumption and distance covered. However, as discussed earlier, this likely reflects limitations in our data coverage rather than truly minimal pilotage activities in these areas.

Italy emerges as the country with the highest fuel consumption per 1,000 nautical miles in our analysis. To explain this finding, we examined the country-specific speed profiles. The data reveals that Italian vessels operate at speeds between 10 knots and 20knots for 46.43% of their total recorded time. This is more than double the average across all countries studied, providing insight into Italy's elevated fuel usage.

To gain a comprehensive understanding of the scale and environmental impact of pilotage activities across Europe, we aggregated the data from all countries in our study. The total distance covered by pilot vessels across all surveyed European countries amounts to 5,521,019.97 nautical miles, which is equivalent to approximately 11,282,606.81 kilometers or approximately 256 circumnavigations of the Earth. This substantial distance underscores the extensive nature of pilotage operations in European waters.

Correspondingly, the total fuel consumption for these operations reaches 24,316.72 tons. These figures highlight the considerable environmental footprint of pilotage activities and underscore the potential for substantial emissions reduction through the implementation of more fuel-efficient practices or technologies such as remote pilotage.

However, it's important to note that these totals likely underestimate the true scale of pilotage activities in Europe, given both the data coverage limitations we've identified for some countries and helicopters and tugs which are also used in pilotage.

Table 3. Distance Made, Total Fuel Consumption, Fuel Consumption per 1000 nm and Average Mission Distance by Country (ordered by t/1000nm)

Country	Total Fuel Consumption (tons)	Distance Made (Nautical Miles)	t/1000nm	Total Missions	Avg Mission Distance (nm)
Italy	49.06	4,316.91	11.36	32	134.90
Georgia	9.18	995.00	9.23	13	76.54
Belgium	1,283.41	148,721.86	8.63	1,025	145.09
Netherlands	4,193.58	578,842.04	7.24	4,084	141.73
Turkey	1,834.24	264,788.26	6.93	1,835	144.30
Ireland	438.72	70,663.10	6.21	604	117.00
Portugal	209.50	34,799.74	6.02	321	108.41
Germany	2,556.15	541,299.28	4.72	5,129	105.54
Morocco	259.72	55,965.65	4.64	597	93.74
Estonia	216.22	48,761.22	4.43	649	75.13
Greece	227.47	51,905.04	4.38	298	174.18
United Kingdom	3,339.01	827,211.39	4.06	5,893	140.37
Poland	355.47	88,062.13	4.04	637	138.25
Latvia	101.56	26,134.08	3.89	330	79.19
France	1,337.81	344,275.10	3.89	2,825	121.88
Sweden	1,179.59	319,564.18	3.69	3,579	89.29
Norway	1,500.80	407,989.53	3.68	3,361	121.43
Denmark	1,887.63	516,975.38	3.65	4,510	114.63
Slovenia	27.94	7,752.88	3.60	56	138.44
Russian Federation	730.72	210,178.79	3.48	2,548	82.49
Faroe Islands	3.31	1,033.60	3.20	10	103.36
Finland	627.77	200,337.37	3.11	2,762	72.53
Malta	165.88	53,520.73	3.10	503	106.44
Bulgaria	37.48	12,375.50	3.00	79	156.65
Spain	1,365.28	481,566.52	2.78	4,081	118.02
Madeira	0.64	236.31	2.71	3	78.77
Ukraine	132.48	53,344.62	2.49	797	66.93
Croatia	72.70	30,976.85	2.35	279	111.03
Montenegro	0.69	337.66	2.05	2	168.83
Lithuania	67.44	37,523.88	1.80	499	75.19
Cyprus	10.30	6,426.81	1.60	76	84.56
Albania	0.90	641.67	1.40	8	80.21
Iceland	3.56	3,313.85	1.07	68	48.73
Romania	68.97	66,517.27	1.04	817	81.42
Gibraltar	21.54	23,665.77	0.91	173	136.80

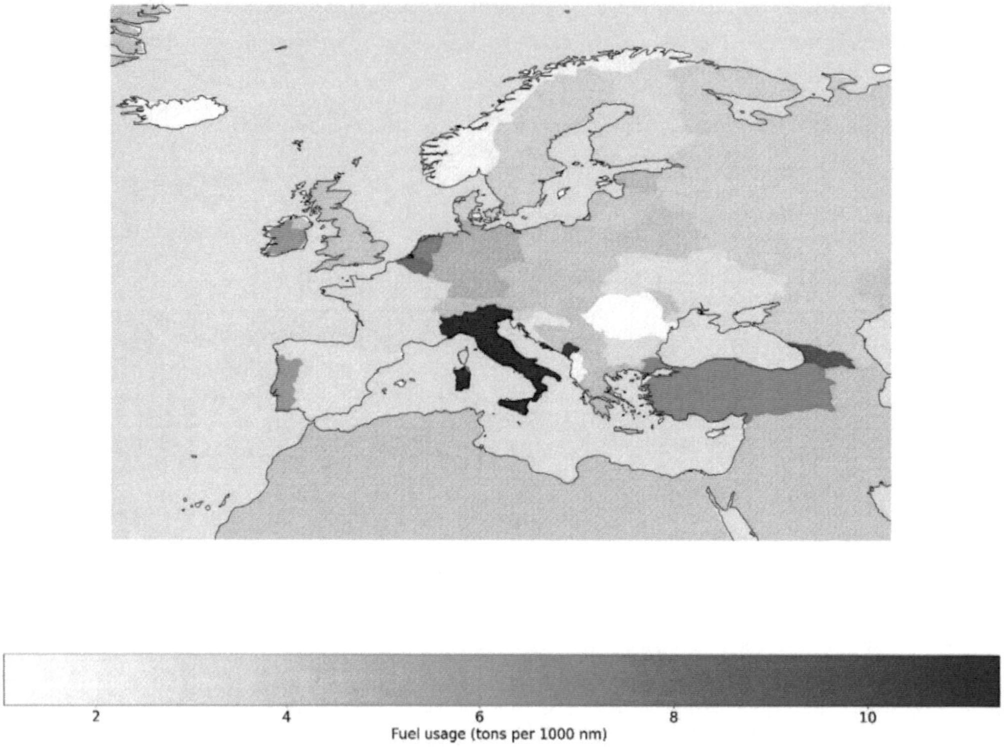

Figure 1: Heat map showing fuel usage (tons) per 1000nm per Country

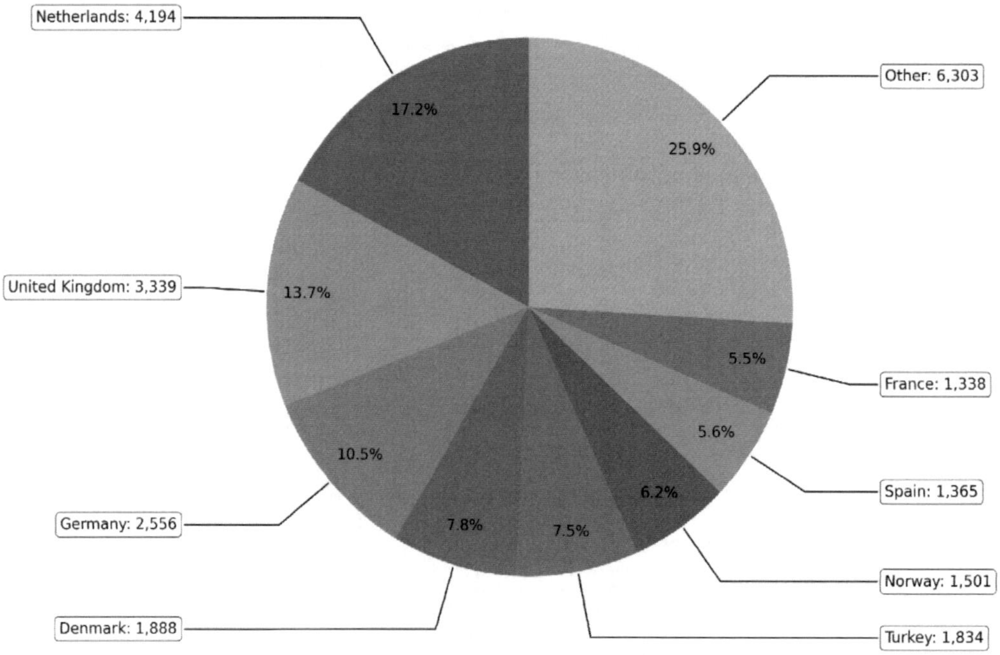

Figure 2: Fuel Consumption (tons) per Country Represented with a Pie Chart

5.4 COMPARISON WITH FORMER STUDIES

For the justification of the data analysis performed in this study, we have compared some of our results in Table 4 with the country-specific results presented in [1]. As [1] consider only the west coast of Sweden and south coast of Norway, we have done the comparison only for Germany and Denmark. Table 4 shows the comparison of several metrics presented in [1] and our paper.

Table 4. Comparison of results between two studies

Country	Metric	Our Results Spanning 2023	Results from [4] Spanning 2019
Denmark	Unique Vessels/MMSI	47	51
Denmark	Total Number of AIS Reports	1,569,663	39,389,917
Denmark	Fuel Consumption (tons)	1,887.63	1,698.29
Denmark	Distance Made (NM)	516,975.38	575,308
Germany	Unique Vessels/MMSI	43	55
Germany	AIS Messages	2,559,953	5,218,007
Germany	Fuel Consumption (tons)	2,556.15	377.31
Germany	Distance Made (NM)	541,299.28	166,022

Key observations are:

- For both countries, we have found slightly fewer unique MMSI than in the first paper.
- Our fuel consumption calculations are higher for both Denmark and Germany, with a particularly large increase for Germany.
- Our distance-made calculations are lower for Denmark but significantly higher for Germany.
- The number of AIS messages in our study is lower for both countries compared to [1].

Some possible reasons that might explain these differences in the results can be given as:

- **Temporal differences:** The four-year gap between the studies (2019 vs. 2023) could account for significant changes in maritime traffic patterns, vessel efficiency, and regulatory environments. This might partially explain the increase in fuel consumption for both countries, particularly the substantial increase for Germany.
- **Data source variation:** The use of different AIS data sources may lead to disparities in coverage, accuracy, and completeness of the data sets. [1] uses AIS data from the Danish Maritime Authority, which has an excellent coverage on Danish waters but substantially lower coverage for other parts of the North Sea, which can explain the difference between fuel consumption and distance made of Germany being higher, on the other hand it also explains why the same parameters are lower for Denmark.
- **Geographical coverage:** Despite focusing on the same countries, there might be differences in the exact geographical areas covered by each study. This could significantly affect the results, especially for distance traveled and fuel consumption, and might explain the contrasting trends observed for Germany and Denmark.
- **Economic and regulatory changes:** Changes in shipping routes, fuel prices, or environmental regulations between 2019 and 2023 could influence vessel behavior and consequently affect fuel consumption and distances traveled. The significant increase in fuel consumption and distance made for Germany in our study might be partially attributed to such changes.

These factors highlight the complexity of maritime data analysis and the importance of considering contextual factors when comparing studies across different time periods and data sources. The substantial differences observed, particularly in fuel consumption calculations and distance made for Germany, underscore the need for careful interpretation of results and the importance of transparent methodology in maritime studies.

6. CONCLUSION

With the advancement of autonomous maritime technologies and the closing of regulatory gaps for remote pilotage applications in first countries, the advent of digital pilotage boarding is on the horizon. Besides safety benefits, this study quantified that there is also a significant potential of fuel savings by remote pilotage, as current pilotage transfer services have been estimated to be around 25,000t within Europe in 2023 resulting in approximately 75,000t carbon dioxide (CO_2) emissions. A reduction would not only have a positive impact on the achievement of European climate targets in terms of CO_2 but would reduce emissions of nitrogen oxides

(NO_x) and sulphur oxides (SO_x). This is in line with the European Union's targets, which aim to reduce overall emissions from shipping by at least 50% by 2050. However, it must be acknowledged that pilot boat emissions are in a total different range than those of maritime shipping in general, which has been around 135 million t in 2022 according to Statista. Nevertheless, further development of this use case is aspired, as Remote Pilotage provides pilots with modern virtual means to execute their tasks to assist the onboard crew in a safer and also more sustainable way.

7. ADDITIONAL INFORMATION

7.1 LIABILITY

Authors are responsible for obtaining security clearance as required. If they so wish authors may add a disclaimer stating that the opinions expressed are solely those of the author. In submitting papers for publication in conference proceedings authors implicitly assign copyright to the Royal Institution of Naval Architects if the paper is published. Where copyright is held elsewhere authors must present evidence of approval to re-publish.

When submitting papers for publication authors must state if the paper has been published before and where or whether it is being considered for publication by some other publisher.

7.2 AUTHOR CONTRIBUTIONS

Conceptualization, H.-C.B.; methodology, H.-C.B. and M.Y.; pilotage process R.G.; autonomous maritime system introduction H.-C.B. and R.G.; supervision, H.-C.B.; AIS extraction and analyses, M.Y.; fuel consumption estimation, N.H.; writing—original draft preparation, M.Y.; writing—review and editing, M.Y. and H.-C.B.; figures: M.Y. introduction, R.G. and H.-C.B.; conclusion and discussion, M.Y. All authors have read and agreed to the published version of the manuscript.

8. ACKNOWLEDGEMENTS

The authors would like to thank Manfred Constapel, for his assistance on gathering the raw AIS data for this study.

9. REFERENCES

1. GRUNDMANN, R., UJKANI, A., WEISHEIT, J., SEPPÄNEN, J., SALOKORPI, M., and BURMEISTER, H C., 'Use Case Remote Pilotage – Technology Overview', *Journal of Physics: Conference Series*, p. 012007, 2023.

2. BURMEISTER, H C., WEISHEIT, J., KÜCHLE, J., BLUHM, L F., and BERGMANN, M., 'An emerging market? The European Maritime Industry's View on Autonomous Maritime Systems: A Survey', *Bd. IOP Conf. Ser.: Mater. Sci. Eng., Nr. in press, p. in press*, 2024.

3. BURMEISTER, H C., BRUHN, W., RØDSETH, Ø J., and PORATHE, T., 'Autonomous Unmanned Merchant Vessel and its Contribution towards the e-Navigation Implementation: The MUNIN Perspective', *International Journal of e-Navigation and Maritime Economy*, Bd. 1, pp. 1-13, 2014.

4. BURMEISTER, H C., CONSTAPEL, M., UGÉ, C., and JAHN, C., 'From Sensors to MASS: Digital Representation of the Perceived Environment enabling Ship Navigation', *IOP Conf. Ser.: Mater. Sci. Eng.,* Bd. 929, 2020.

5. KOCH, P., STACH, T., CONSTAPEL, M., BURMEISTER, H C., GILLERT, A., and VAHL, M., 'Study on the Evaluation of Algorithms for Nautical Applications (EANA)', 2024: BSH.

6. BURMEISTER, H C., and CONSTAPEL, M., 'Autonomous Collision Avoidance at Sea: A Survey', *Front. Robot. AI*, Bd. 8, 2021.

7. STACH, T., KINKEL, Y., CONSTAPEL, M., and BURMEISTER, H C., 'Maritime Anomaly Detection for Vessel Traffic Services: A Survey', *Journal of Marine Science and Engineering, Bd. 6*, 2023.

8. VEITCH, E., and ALSOS, O A., 'A systematic review of human-AI interaction in autonomous ship systems', *Safety Science, Bd.* 152, 2022.

9. VAN DEN OEVER, F., FJELD, M., and SÆTREVIK, B., 'A Systematic Literature Review of Augmented Reality for Maritime Collaboration', *International Journal of Human–Computer Interaction , Bd.* 15, 2024.

10. UJKANI, A., HOHNRATH, P., GRUNDMANN, R., and BURMEISTER, H C., 'Enhancing Maritime Navigation with Mixed Reality: Assessing Remote Pilotage Concepts and Technologies by In Situ Testing', *Journal of Marine Science and Engineering 12*, 2024.

11. International Maritime Organization, 'Revised Guidelines for the Onboard Operational Use of Shipborne Automatic Identification Systems (AIS)', 2015.

12. BLOUNT, D L A M J., 'Resistance characteristics of semi-displacement mega yacht hull forms', Trans. RINA, Intl. J. Small Craft Tech, Bd. 151, 2009.

13. IMO, 'Pilot Transfer Arrangements: Resolution A.1045(27)', [Online].

14. I. M. P. Association, 'IMPA Guidance on the Master - Pilot Exchange (MPX)'

15. SEPPÄNEN, J., and SALOKORPI, M., 'Critical Ship Data in Remote Pilotage: Sea4Value', in *Sea4Value -- Future Fairway Navigation, https://www.dimecc.com/dimecc-services/future-fairway-navigation/*, pp. 60-77, 2022.

16. World Food Programme, 'https://public.opendatasoft.com/', 26 April 2019. [Online]. Available: https://public.opendatasoft.com/explore/dataset/world-administrative-boundaries/information/. [Zugriff am 30 September 2024].

17. International Association of Marine Aids to Navigation and Lighthouse Authorities, 'IALA Guidelines 1082 An Overview of AIS', 2016.

10. AUTHORS BIOGRAPHY

Mert Yildiz is an applied data scientist, focusing on combining machine learning methods to elevate existing AIS analysis capabilities. He is a mechanical engineer with MSc. in Mechatronics.

Nils Hagemeister is research associate, focusing on performance assessments of cargo ships. He is an engineer with M.Eng. in Naval Architecture.

Robert Grundman is team leader "Nautical Solutions", focusing on maritime human-centered engineering, mixed reality applications and ship handling simulation-based validations. He has a captain's license as well as a diploma in Nautical science and Maritime traffic.

Hans-Christoph Burmeister is Head of Department "Sea Traffic and Nautical Solution", focusing on nautical innovations, autonomous maritime technologies and navigational assessments. He is an industrial engineer with a diploma in mechanics and informatics.

CONCEPT DESIGN ON A LARGE UNMANNED SURFACE VEHICLE (LUSV)

P Curtolo, Italian Navy, Italy
T Russo and **G A Pacifico,** Italian General Navy Staff – Ship Design Office, Italy
E Begovic and **S Mancini,** Department of Industrial Engineering – University of Naples Federico II, Italy

SUMMARY

The present paper discusses the concept design of a Large Unmanned Surface Ship (LUSV) with an overall length of seventy meters and a maximum speed of 20 knots, with a similar Concept of Operations (CONOPS) as a light patrol vessel. The ship should be operated without seafarers on board and be remotely controlled by a land-based controller room, with a level of autonomy between three or four, according to the IMO MASS Code. The study starts with the definition of the general arrangement. Taking into consideration the absence of the crew. spaces such as mess halls, kitchens, and accommodations are no longer necessary. However, additional spaces are required to fulfil functions previously carried out by crew members. To ensure an equivalent level of safety to a manned vehicle, redundant systems and sub-divisions of vital systems and spaces are included in the design. Additionally, improved fault diagnosis devices, enhanced processes for data collection and transmission, and automation of the ship's apparatus are also incorporated. The concept design addresses technical aspects such as manoeuvring and seakeeping capabilities, structural scantling, and stability to complete the concept design of the Large Unmanned Surface Ship, identifying the pros and cons of the proposed solutions compared to a traditional patrol vessel solution.

NOMENCLATURE

A	Ship manoeuvring advance (m)
A_E	Propeller expanded area (m^2)
A_O	Propeller disc area (m^2)
A_P	Longitudinal projected area (m^2)
A_T	Transverse projected area (m^2)
c	Appendage mean chord (m)
C_F	Frictional coefficient (ITTC-57)
d	Shaft diameter (m)
D	Propeller diameter (m)
H_S	Significant wave height (m)
k_s	Allowance value (μm)
L	Ship waterline length
L_{SH}	Shaft overall length (m)
P	Propeller pitch (m)
Re	Reynolds number
S	Hull surface (m^2)
$S_{BILGE\ KEEL}$	Bilge keel surface (m^2)
t	Appendage thickness (m)
T	Wave period (s)
v	Ship speed (m s^{-1})
V_R	Apparent wind velocity (m s^{-1})
ε	Relative angle between shaft and flow (rad)
ρ	Density of water (kg m^{-3})

ACRONYMS

AAW	Anti-Air Warfare
ANEP	Allied Naval Engineering Publication
ASW	Anti-Submarine Warfare
ASuW	Anti-Surface Warfare
CBRN	Chemical, Biological, Radiological, and Nuclear
CONOPS	Concept of Operations
DMO	Distributed Maritime Operations
EAR	Expanded Area Ratio
EEZ	Exclusive Economic Zone

HM	Harbour Manoeuvring
HVAC	Heating, ventilation and air conditioning
IFEP	Integrated Full Electric Propulsion
IR	Infrared
ITTC	International Towing Tank Conference
LARS	Launch and Recovery System
LIDAR	Light Detection and Ranging
LUSV	Large Unmanned Surface Vehicle
MASS	Maritime Autonomous Surface Ships
MCM	Mine Countermeasure
MIO	Maritime Interdiction Operation
MMG	Maneuvering Modelling Group
MSC	Maritime Safety Committee
OPAREA	operative area
RHIB	Rigid-Hull Inflatable Boat
RMS	Root mean square
RSE	Regulatory Scoping Exercise
SAAM ESD	Surface- Anti Air Missile – Extended Self Defence
SAR	Search and Rescue
SLOC	Sea Lines of Communication
STANAG	Standardization Agreement
SUW	Surface Warfare
TAP	Transit and Patrol
UAV	Unmanned Aerial Vehicle
USV	Unmanned Surface Vehicle
VI	Vehicle Interaction
VLS	Vertical Launching System
WBS	Work Breakdown Structure

1. INTRODUCTION

In recent years, the arsenals of all armed forces have been enriched with unmanned vehicles, characterized by the absence of crews and pilots on board. The aerial domain has undoubtedly played a main role in the evolution of drones, both for intelligence gathering and target engagement. However, there is a growing interest in the maritime domain, as seen in the Ukrainian attacks which were carried out by a flotilla of USVs on October 29th, 2022.

Regulatory frameworks are recognizing this trend. The IMO accepted to undertake a Regulatory Scoping Exercise (RSE) on MASS (Maritime Autonomous Surface Ships) at the 98th session of the Maritime Safety Committee (MSC 98) in February 2017. The RSE was carried out through subsequent sessions investigating safety, security, environmental, liability and personnel concerns of unmanned technology. The scoping exercise was concluded at 108th session in July 2021. Following the RSE, the aim is to develop a goal-based code for MASS operations to take effect in 2025 [1] [2] [3].

The IMO has classified MASS into four classes for the RSE depending on the degree of autonomy (from 1 being the lowest level up to 4-completely autonomous vehicle) [4]. On the other side, USV can be classified based upon their size. According to US Navy's USV systems vision, there are very small USV up to 7 meters, small USV between 7 and 12 meters, medium USV from 12 and 50 meters and large USV above 50 meters [5]. A review of thirty-eight existing vessels and conceptual studies in technical literature, encompassing both civil and military application and belonging to nine different nations, revealed a wide variety of small USVs, while there are a few examples of large-scale vehicles. USVs maximum speed ranges from few knots up to 60 knots.

The main tasks of military USVs are surveillance, search and rescue (SAR), Mine Counter Measure warfare (MCM), Anti-Submarine Warfare (ASW), Anti-Air Warfare (AAW), Anti-Surface Warfare (ASuW) whereas civil USVs are used for oceanographic and environmental purposes, as well as for offshore platform logistic support.

The aim of the present paper is to analyse the concept design of a Large Unmanned Vehicle. The objectives are to investigate possible alterations of ship's arrangements, analyse alternative solutions to achieve an equivalent level of safety and evaluate technical aspects such as manoeuvrability, seakeeping, hull scantling, and stability.

2. CASE STUDY

The reference ship of this study is a light patrol vessel. The methodology consists of removing the crew from the reference ship and analyse how ship's design and operations change maintaining similar CONOPS.

The main requirements of the ship are safeguard of the national jurisdiction and coastal waters, surveillance of the Exclusive Economic Zone (EEZ), prevention of smuggling of peoples, drugs and weapons, Maritime Interdiction Operations (MIOs) and sea denial, anti-pollution activities, monitoring of offshore structures and of fisheries.

LUSV must perform all the tasks with a degree of autonomy between three and four of IMO MASS classification.

The operative area (OPAREA) of the ship is the enlarged Mediterranean, i.e. the area between Gibraltar and the Gulf of Aden including the Middle East and Central Africa [6].

The ship must ensure safe and secure navigation even in case of failure occurrence choosing robust and reliable technologies, homogeneous and heterogeneous redundancy and backup systems minimizing the impact on operations. LUSV can remain on-station for months with no host-nation support [7] [8].

The propulsion system must be redundant and divided into compartments to reduce mechanical loss issues. In addition, the ship must implement advanced encryption and security systems to prevent potential cyber threats. Data must be collected and stored in independent data centres where they will be processed and sent to a remote-control centre or control algorithm.

In case of emergency, the ship must be reached and recovered either with landing crafts (Rigid-Hull Inflatable Boat, RHIB) or with a medium-sized helicopter releasing a boarding team.

LUSV must be able to deploy aerial, underwater, and surface vehicles.

Ship's capabilities can be summarized as follows:

- Capability to operate with a medium-sized helicopter, such as the NH-90;
- Capability to recover and deploy UAVs;
- Navigation, detection, and communication capabilities comparable with a traditional light patrol vessel;
- Two vertical launchers (VLS, Vertical Launching System) with 8 cells each, with SAAM ESD (Surface- Anti Air Missile – Extended Self Defence) Missile System and ASTER 15 and 30 missiles;
- Aft LARS for a 9.35-meter rigid-hulled inflatable boat type Hurricane or USV of the same size;
- Launch and recovery system for UUVs;
- Diesel-electric propulsion type (IFEP, integrated full electric propulsion) with completely separate propulsion lines and fixed-pitch propellers;
- Adequate compartmentalization and redundancy of power generation. The ship should be provided with three independent and separate engine rooms per shaft line;
- An emergency control room;
- A bow thruster with take-home capabilities in case of loss of ship propulsion system;
- Maneuvering capabilities comparable to traditional units. No special mooring capabilities are required as the unit can be towed once it reaches port;
- A maximum speed of 20 knots;
- The unit must have greater autonomy than that of a light patrol vessel for cruising speed, as there are no longer logistical limitations related to operators.

3. RESULTS & DISCUSSION CONCEPT DESIGN

The main characteristics of the ship must fulfil the set of requirements defined in the previous section. All criteria are analysed to estimate the ship overall dimensions.

The recovery capability is one of the core features of the ship. LUSV, operating with both helicopters and UAV, is equipped with a flight deck compliance with Italian Navy regulations [9]. The flight deck is twenty-two meters long and fourteen meters wide.

Secondly, the ship has an adequate superstructure that combines navigational and communication capabilities. The ship includes a foremast, which houses surface search radar, navigational radars, and other sensors. The superstructure also houses cameras, LIDAR, infrared (IR), acoustic sensors, and electronic warfare suites. Sufficient situational awareness requires that all detected objects or conditions be classified, and any change in state be determined so that feasible future states can be established [10] [11]. It is supposed twenty-seven meters-long overall superstructure.

The combat system includes a Vertical Launching System (VLS) in the fore part of the ship. The VLS has an eight-cell dual-packed missile configuration. The overall longitudinal footprint of the VLS is eight meters long. The minimum distance from the bow is ten meters long and the clearance from the superstructure is about one meter.

The aft area of the LUSV consists of a system to launch and recover (LARS) traditional surface assault craft as well as Unmanned Surface Vehicles (USVs). Launch and recovery operations must be carried out completely autonomously, without human intervention. The overall footprint of the LARS is 23.5 meters long. The Integrated Full-Electric propulsion plant (IFEP) is designed with an independent-bus diesel-electric propulsion system consisting of twin-screw propellers driven through shaft lines. The shaft lines must be separated by a longitudinal bulkhead for safety reasons. Each independent-engine room incorporates auxiliary enhancements, such as fuel tanks, lubricating oil, wicks, and pumps.

There is a reduction of on-board spaces due to the elimination of accommodations, messes and quarters compared to a traditional manned vessel. The ship consists of three decks, two of which are dedicated to machinery allocation and one for ship apparatus, data centre and automation distribution, and a bilge. Each deck maintains a 2.5-meters deck-to-deck distance for maintenance accessibility while the bilge is one meter high. It results an overall length of 70 meters, a breadth of 14 meters, and a depth of 8.5 meters.

The displacement of the ship was determined using scaling factors estimated for similar vessels. The scaling factors used were power ratio and cubical number ratio. The deadweight was calculated using empirical equations available in literature [12]. The LUSV has a displacement of 2000 tons and draught of 3.75 meters. The hull form was modelled from existing similar ships. Table 1 outlines the main characteristics. Additionally, Figure 2 depicts the hull form and 3D view of the vessel.

Table 1. Main characteristics of the LUSV.

Specification		Unit	Value
L_{OA}	Length Over All	m	69.5
L_{WL}	Waterline Length	m	68.1
B_{MAX}	Maximum Beam	m	13.8
B_{WL}	Beam at Waterline	m	14.1
D	Depth	m	8.5
T	Draft	m	3.75
Δ	Displacement	t	2000
C_B	Block Coefficient	-	0.555

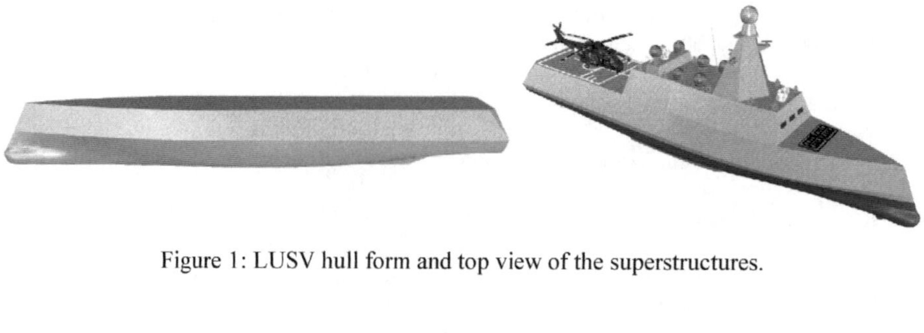

Figure 1: LUSV hull form and top view of the superstructures.

3.1 PROPULSION DESIGN

The LUSV is designed with an IFEP system. The power generation system consists of six diesel engines arranged in separate and watertight spaces for redundancy purposes. The diesel generators drive two electric motors with two shafts and two fixed-pitch propellers (FPP). The diesel-electric propulsion plant has significant advantages. These include reduced ship vulnerability, high operational flexibility and decreased noise and vibration.

The total resistance and effective power of the ship were assessed using the Fung statistical method [13]. Additional resistance components were estimated using various established methods, i.e. Hoerner's formulas [14] were applied for the resistance of appendages such as rudders, struts, shaft lines, and bilge keels:

$$D_{shaft} = \frac{1}{2} \bullet \varrho \bullet L_{SH} \bullet d \bullet v^2 \bullet \left(1.1 \bullet sin^3\varepsilon + \pi \bullet C_F\right)$$

$$D_{rudder-struts} = \frac{1}{2} \bullet \varrho \bullet A_p \bullet v^2 \bullet 2 \bullet \left(C_F + 0.0008\right) \bullet \left(1.2 \bullet \frac{t}{c} + 1\right)$$

$$R_{T-BILGE\ KEEL} = \frac{1}{2} \bullet \varrho \bullet \left(S + S_{bilge\ keels}\right) \bullet v^2 \bullet C_T$$

Carlton's procedure [15] was employed to account the impact of fouling:

$$\Delta C_{F-FOULING} = 0.044 \bullet \left[\left(\frac{k_s}{L}\right)^{\frac{1}{3}} - 10 \bullet Re^{-\frac{1}{3}}\right] + 0.000125$$

Hughes's empirical formula was used for air and wind resistance [3]:

$$R_{WIND} = 0.734 \bullet A_T \bullet V_R^2$$

Added resistance was estimated by means of literature formulas. The propulsive factors were estimated considering ships with similar characteristics.

The propellers were optimized at cruise speed of twelve knots and two five-blades Wageningen BB propellers were selected. They have a diameter (D) of 2.3 m, a pitch-to-diameter (P/D) ratio of 1.3 and an expanded-to-disc-area ratio (EAR) of 0.8. The propeller characteristics were verified for cavitation at the maximum speed using the Burril's diagram.

The ship is equipped with two electric motor and six diesel generators. Each electric motor can produce 4000 kW at 300 rpm. The electrical load analysis was determined for each operational condition using literature formulas [16]. Considering efficiencies and redundancy requirements, the total installed power is around 14000 kW.

3.2 GENERAL ARRANGEMENT

The distribution of spaces is strictly connected with the requirements of the ship. The ship is divided into 5 decks. There are technical rooms in the lower part of the ship whereas data centres and armament are allocated in the upper part.

Decks 3 and 4 house the six diesel-generator rooms, the two electric motor rooms, VLS equipment, the dock area with two steering gear technical rooms, and auxiliaries. The technical spaces are separated by a longitudinal watertight bulkhead and there are no passageways connecting each room. The ship has vertical openings for maintenance accessibility. The mooring area, equipment and arrangement were downsized and integrated with an emergency anchor due to the absence of personnel on board. The ship will be towed to the berth once in port. Sewage, fresh and sanitary water tanks, larder, and fridge rooms are no longer required. The engine rooms are equipped with all propulsion auxiliaries such as independent fuel storages and service tanks, sludge, oil tanks, exhaust, and inlet pipes.

The main data centres are located on deck 1, which is the main deck, and 2. The LUSV must be able to store, analyse and transmit information collected by all on-board sensors. The electric panel rooms are separated and contain the busses, switchboards, and batteries. The batteries can prevent brownouts and blackouts. There are two heating, ventilation, and air conditioning (HVAC) rooms necessary for refrigerating ship apparatus. The ship is not equipped with chemical, biological, radiological, and nuclear (CBRN) defence system. The main deck is equipped with a flight deck and two hangars for recovering UAVs. There is also an emergency bridge accessible via a passageway. In case of emergencies, a team could be deployed through the flight deck to take control of the ship. Mess, accommodations, fitness, galley, quarters and hospital rooms were not included.

The superstructure allocates spaces for communication capabilities, refrigeration of apparatus and weapon system. There is a passageway on deck 01 for emergency purposes.

The ship is provided with eight transverse watertight bulkheads extending from the baseline to the main deck. The longitudinal watertight bulkhead separates all technical spaces from the baseline to the main deck as shown in Figure 3.

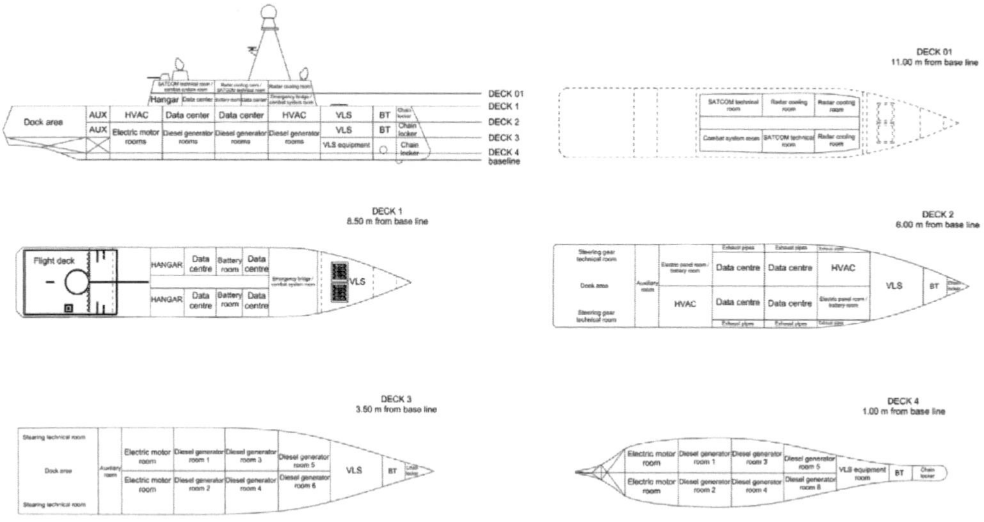

Figure 2: General arrangements of the LUSV.

3.3 SHIP STABILITY & HULL SCANTLING

The ship is compliant with RINAMIL [17] requirements for stability and structural scantling. Three loading conditions were considered: full load, operational condition, and minimum operation condition both for intact and damaged stability were considered. The intact stability criteria were defined in compliance with RINAMIL standards, taking into account the requirement for the righting arms curve, the effects of free surfaces and the severe wind and rolling effects. For damage cases, a longitudinal extension of two adjacent compartments was considered. For each loading condition and damage case all applicable criteria must be complied with taking into account the effects of free surfaces. The analysis was conducted by using the software Maxsurf Stability 2023.

The midship section was analysed for structural scantling. The main characteristics of this section were evaluated. It was chosen a DH36 hull stell with a minimum yield stress of 355 N/mm^2, an ordinary stiffener spacing of 600 mm and a primary supporting member spacing of 1800 mm. The design loads were defined in accordance with the RINAMIL.

3.4 SEAKEEPING

The operability of a military ship is strictly connected to its loading condition and the missions a vessel must perform. An autonomous vehicle's operability is not influenced by the crew seasickness and fatigue, but it can be limited by electronic equipment, such as weaponry and antennas, which are sensitive to movements.

Seakeeping analyses were conducted in compliance with Standardization Agreement (STANAG) 4154 [18] and Allied Naval Engineering Publication (ANEP) 4154 [19] requirements considering full load condition.

The regulations propose conventional tasks. Three missions have been analysed according to the ship's deployment: Transit and Patrol (TAP), Surface Warfare (SUW), and Vehicle Interaction (VI). Each mission has specified speed values, reference points and seakeeping criteria. Sea states were defined by (ITTC/JONSWAP)) spectrum with a wave height (H_S) of 1.89 meters and peak period (T_P) of 8 seconds. The analysis was conducted in the frequency domain by using the software ShipX 4.0. All values refer to the root-mean square (RMS). Calculation parameters are summarized in Table 4. Additionally, the reference points are shown in Figure 5.

Table 2. Seakeeping missions and criteria.

Type of mission	Ship speed	Sea state	Criteria	Reference point
Transit And Patrol (TAP)	12 knots	SS4	24/hr	Sonar emergence
			7.5°	Sonar roll
			2.5°	Sonar pitch
			30/hr	Wetness (at the VLS)
			20/hr	Slamming (at the sonar dome)
			90/hr	Propeller emergence
Surface Warfare (SUW)	20 knots	SS4	1.5°	VLS Pitch
			0.8°	VLS Yaw
			0.15g	VLS Long. Acc.
			0.35g	VLS Tran. Acc.
			0.3g	VLS Vert. Acc.
			24/hr	Sonar emergence
			3.8°	Gun Roll
			0.5 m/s	Gun Vert. Vel.
			30/hr	Wetness (at the VLS)
			20/hr	Slamming (at the sonar dome)
			90/hr	Propeller emergence
Vehicle interaction (VI)	8 knots	SS4	2.5°	NH-90 Launch/Recovery Roll
			1.25°	RHIB Pitch Angle
			0.1g	RHIB Vert. and Lat. Acceleration

Figure 3: Reference points for the seakeeping analysis.

The analysis shows an improvement in ship seakeeping performance connected to the absence of the "human factor". The results highlight that ship operability could be extended to a higher sea state (SS5) for TAP, while the ship does not satisfy all criteria for SUW and VI. The main reason is the demanding reference environmental conditions and speeds used during the investigation and the too restrictive thresholds for motion criteria. The operability diagrams are shown in Figure 6.

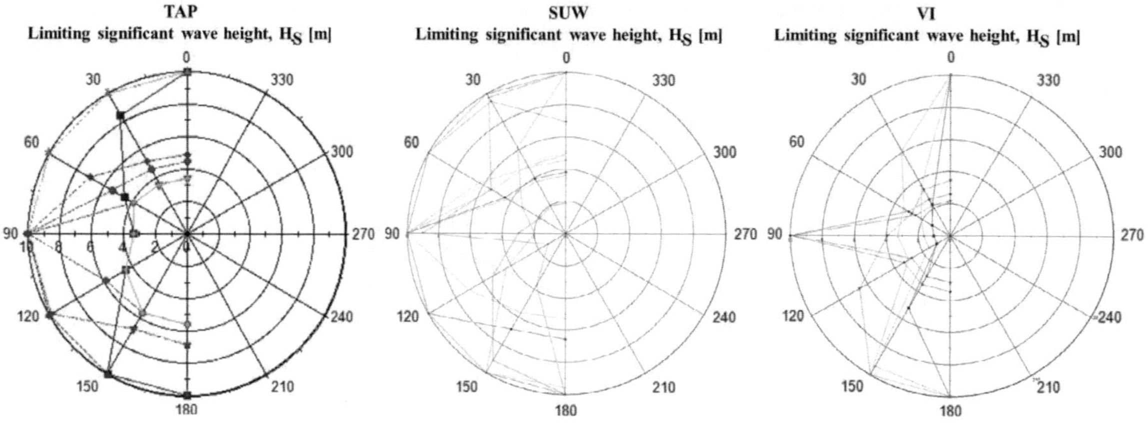

Figure 4: Operability diagrams.

3.5 SHIP MANEUVERING

In the ship design process, manoeuvrability is typically only considered after all other aspects have been defined. At that stage, it is difficult to make any significant changes to the design. This paper assesses ship manoeuvring according to Manoeuvring Modelling Group (MMG) theory. The regulatory framework consists of ANEP 70 [20] guidelines and IMO MSC 137 (76) [21] standards. Following regulations, Transit and Patrol and Harbour Manoeuvring missions were considered. For each, it was defined ship parameters and criteria. The study utilized the MANSIM software, which implements the MMG [22] theory. Table 7 summarizes the input parameters for manoeuvring, while Figure 8 illustrates the simulated manoeuvring tests.

Table 3. Manoeuvring missions and criteria.

Mission	Speed	Criteria	STANAG	IMO	Results
Transit And Patrol (TAP)	12 knots	Course keeping	+/- 4°	N.A.	0.0
		Tactical diameter	3 of ship length	5 of ship length	2.0 of ship length
		Initial turning	10 sec	27.6 sec	8.0 sec
		Yaw check	7 sec	N.A.	10.0 sec
		Yaw check first o.a.	N.A.	11°	9.96°
		Yaw check second o.a.	N.A.	26°	15.68°
		Yaw check first o.a.	N.A.	25°	17.15°
		Advance	N.A.	4.5 A/L	2.03 A/L
Harbour Maneuvering (HM)	4 knots	Course keeping	+/- 4°	N.A.	0.0
		Tactical diameter	2 of ship length	5 of ship length	1.8 of ship length
		Initial turning	10 sec	83 sec	20.0 sec
		Yaw check o.t.	10 sec	N.A.	18.0 sec
		Yaw check first o.a.	N.A.	20°	6.64°
		Yaw check second o.a.	N.A.	40°	11.75°
		Yaw check first o.a.	N.A.	25°	10.45°
		Advance	N.A.	4.5 A/L	1.85 A/L

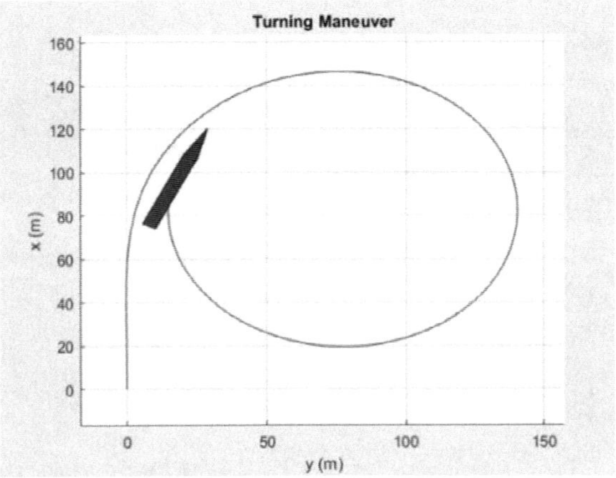

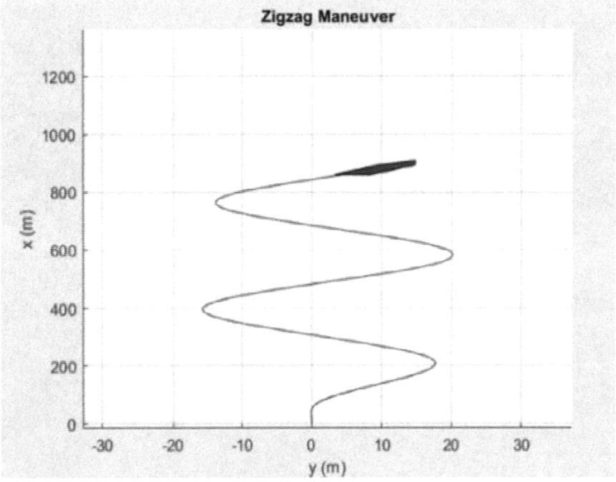

Figure 5: Turning manoeuvre and zig-zag manoeuvre simulated tests.

All IMO MSC criteria were satisfied while some ANEP 70 were not, as were considered too demanding. The unverified criteria were yaw check for both conditions and initial turning for the slow speed condition. Additionally, the MMG theory considers the planar motion assumption which simplify the analysis (no roll motion considered). Further analysis will be requested using more advanced methodologies to assess the manoeuvring performances, especially at low speeds.

4. CONCLUSIONS

In recent years, the interest in autonomous vehicles has grown due to the innovations and advantages they can bring in armed conflicts. This paper focuses on the development of a large, unmanned vehicle, assessing the impact on the ship's layout and analysing possible solution to ensure an equivalent level of safety. After determining the main dimensions, the hull shape was chosen, and the displacement and power prediction were estimated. Finally, technical aspects were analysed.

This concept design highlights the feasibility of developing a large, unmanned vehicle. showing the strength points and the critical issues. Autonomous vehicles can expand the operational area of all maritime nations, in accordance with the distributed maritime operations (DMO) doctrine [23]. Furthermore, USVs reduce the risks associated with ship deployment in wartime scenarios due to the absence of humans.

It can provide economic benefits by reducing the overall length of a ship compared to an equivalent traditional manned unit. However, there may be an increase in costs to ensure an equivalent level of safety in terms of redundancy and independence of all vital systems. The resulting loss or profit must be assessed with further analysis. Lastly, USVs have better operability in terms of seakeeping, which is not limited by the crew's seasickness and fatigue.

On the other hand, threats such as piracy and cyber-attacks can be critical for autonomous ships. USVs must have secure closure systems for all external access. Additionally, redundant and enhanced encryption communication systems must be completely separated to protect against cyber-attacks. If this is not possible, it must be considered to implement a higher degree of autonomy by using algorithms to make the ship independent from external communication. At the same time, all communication infrastructure must be enhanced to meet the data size requirements in high seas, evaluating the possibility of installing pre- or post-data processing systems.

5. ACKNOWLEDGEMENTS

The authors thank all the members of the IT Navy Ship Design Office for the support during the project.

6. REFERENCES

1. Maritime Safety Committee (MSC), 'Interim Guidelines for Mass Trials', *MSC.1/Circ. 1604*, 2019.

2. Maritime Safety Committee MSC, 'Development of A Goal-Based Instrument for Maritime Autonomous Surface Ships (MASS)', *MSC 105*, 2022.

3. Intersessional Working Group on Maritime Autonomous Surface Ships (MASS) (MSC/ISWG/MASS), Report of the Working Group, *MSC 107*, 2023.

4. Maritime Safety Committee MSC, 'Outcome of the Regulatory Scoping Exercise for the Use of Maritime Autonomous Surface Ship', *MSC.1/Circ. 1638*, 2021.

5. Congressional Research Service (CRS), 'Navy Large Unmanned Surface and Undersea Vehicles: Background and Issues for Congress', *CRS Report*, 2024.

6. DI CECCO, V., 'A "great Middle East" or an "enlarged Mediterranean" (in Italian)', *Italian Information of Defence magazine*, 2004.

7. ABS, 'Autonomous Vessels Whitepaper', 2022.

8. Group Technology and Research, 'Remote-controlled and autonomous ships in the maritime industry', *DNV*, 2018.

9. Italian Navy Commanding Officers, 'SMM-AER-003', 2021.

10. ABS, 'Requirements for Autonomous and Remote Control Functions', 2022.

11. DNV, 'Autonomous and remotely operated ships', *Class guideline — DNV-CG-0264*, 2021.

12. PAPANIKOLAOU, A., 'Ship Design: Methodologies of Preliminary Design', *Springer*, 2014.

13. FUNG, S C., and LEIBMAN, L., 'Statistically-based speed-dependent powering predictions for high-speed transom stern hull forms', *FAST '95: Third International Conference on Fast Sea Transportation*, 1995.

14. LEWIS, E V., 'Principles of Naval Architecture', *The Society of Naval Architects and Marine Engineers*, 1988.

15. CARLTON, J S., 'Marine propellers and propulsion', *Elsevier Ltd.*, 2007.

16. Brix, 'Manoeuvring Technical Manual', *Seehafen Verlag*, 1993.

17. RINA, 'Rules for the classification of naval ships', 2017.

18. Military Agency for Standardisation (MAS), *'Common procedures for seakeeping in the ship design process', Standardisation Agreement (STANAG) 4154*, 2013.

19. Nato Standardisation Office, 'Common procedures for seakeeping in the ship design process', *Allied Naval Engineering Pubblication (ANEP) 4154*, 2013.

20. Naval Group 6 specialist team on naval ship manoeuvrability, 'Guidance for naval surface ships mission oriented manoeuvring requirements', *Allied Naval Engineering Pubblication (ANEP) 70*, 2003.

21. Maritime Safety Committee (MSC), 'Standard for ship Manoeuvrability', *MSC 137(76)*, 2002.

22. YOSHIMURA, Y., 'Mathematical Model for Manoeuvring Ship Motion (MMG Model)', *Workshop on Mathematical Models for Operations involving Ship-Ship Interaction*, 2005.

23. BARBERINI, P., 'The development of unmanned maritime systems (in Italian)', *Italian Maritime magazine*, 2022.

7. AUTHORS BIOGRAPHY

P Curtolo holds the current position of Lieutenant (jg) at Italian Navy.

T Russo holds the current position of Captain at Italian Navy.

G A Pacifico holds the current position of Lieutenant Commander at Italian Navy.

E Begovic holds the position of Associate Professor of Ship Seakeeping and she is the Coordinator of the Master Mundus Erasmus Joint Master Degree Sustainable Ship and Shipping 4.0 (SEAS 4.0) from 2020 and responsible for International agreements and Erasmus program for the Department of Industrial Engineering. Her main research interests are concerning High Speed and Planing Craft Seakeeping, Design and Optimization, Experimental Hydrodynamics, and 2nd Generation Intact Stability Criteria. Chairman of ISSC Specialist Committee on Special Craft (2022-); member of Stability Research and Development Committee (2019-), editor of High-Speed Marine Vehicles 2020 and 2023 symposium proceedings and JMSE Special Issue "Stability and Seakeeping of Marine Vessels". Ermina Begovic is the author of more than 100 scientific papers published in journals or presented at international conferences.

S Mancini holds the current position of Associate Professor at University of Naples "Federico II".
He is also working as a consultant for FORCE Technology/DMI, where he was the Senior Team Leader of the Numerical Team at the Department of Hydro and Aerodynamics for four years. For fifteen years, he served as a Technical Officer of the Italian Navy, where he had periods of experience on board and acted as a Project Manager at the Navy General Staff - Ship Design Office. He is the author of more than sixty papers published in international journals and conference proceedings and is editorial board member of several international journals (Journal of Marine Science and Engineering, Journal of Marine Science and Applications, Brodogradnja - An International Journal of Naval Architecture and Ocean Engineering for Research and Development, and Cogent Engineering).

MUM - AN EXTRA LARGE UNMANNED UNDERWATER VEHICLE (XLUUV)

P Schön and **M Riesner**, thyssenkrupp Marine Systems; Germany

ABSTRACT

The joint research project "MUM – Large Modifiable Underwater Mothership" develops a family of highly modular, uncrewed underwater vehicles. The MUM system allows a mission-dependent module assembly to fulfil a wide spectrum of underwater tasks. A full-scale demonstrator of approx. 25 m length is being produced. The paper presents the project's goals and selected results of the extra-large uncrewed underwater vehicle's (XLUUV) development. It presents challenges the team faced and highlights the differences between designing one-of-a-kind surface vessels and an adaptable XLUUV concerning autonomy and regulatory aspects. As an XLUUV, the MUM must not only field highly automated surface navigation but also underwater navigation. The paper provides insight into how hazard and risk assessments guided the engineering and how early regulatory stakeholder engagement is one of the project's success factors. The paper concludes with an outlook to the verification and validation phase of demonstrator.

1. INTRODUCTION

Work and live at sea is permanently affected by meteorological conditions. Illustrations often assume fair weather conditions, which, however, usually only occur on a few days a year. Modern offshore supply vessels (OSV) are specifically designated for the operation in offshore environmental conditions and optimized based on advanced technical support systems to assist and increase operability. To enable the operation in wind, wave and currents, OSV are equipped with a dynamic positioning system (DPS). The DPS counteracts against wave-, wind- and current-induces forces by generating thruster-induced counterforces. Computations e.g. by Mauro and Nabergoj (2022) or Veer and Gachet (2011) shows that dynamic positioning of OSVs might be applicable up to wind speed of more than 60 kn and waves with a significant wave height of 15 m, when wind and waves encounter from bow or stern direction. In bow-quartering and side wind and wave direction, the DP capabilities reduce significantly. Such DP capabilities may still illustrate that environmental conditions do only marginally effect the operability of OSVs. However, such dynamic positioning capability investigations are usually performed considering time average forces as well as time average ship motions. First order wave-induced ship motions, such as heave, pitch and roll, are not considered during such predictions, however, they play a significant role for the operability of such vessels. For example, launching and recovering of remotely operated vehicles (ROV) or autonomous underwater vehicles (AUV) are strongly affected by first order ship motions. Exceeding significant wave heights of 2.5 m already have a considerable impact on the operability. Sea state statistics such as Hogben et al. (1986), Söding (2001) or BMT Agross BV (2013) include wave probability data for almost every sea area, showing that significant wave heights higher than 2.5 m occur frequently at typical sea areas with offshore industries, especially during winter season. That's why non-productive costs, such as waiting for better weather conditions, are about 30% of the total operating vessel costs, Dick (2023).

The research project MUM2 has the goal to develop a full size demonstrator of an autonomous extra large unmanned underwater vehicle with a length of 26 meters. Based on the large, and for UUVs unusual, main dimensions, MUM is capable of replacing the usually required OSVs fully. MUM can operate autonomously or remotely controlled via a communication buoy. Replacing surface vessels by underwater vehicles (significantly less affected by waves and wind) and operating autonomously strongly increase the operability time per year and decreases personnel costs enormously.

A further design feature of MUM (Modifiable Underwater Mothership) is its modularity. The vehicle is divided into basic modules, such as a power supply module and dedicated individual mission modules. The basic modules are required for every MUM operation. The mission modules can be interchanged to fulfil the customized requirement of the individual vehicle application and could also include customized enhancements of range and speed. This technique gives MUM a wide range of applications. Furthermore, each component of the vehicle can be dismounted to a size of less or equal to a 20 foot container. This allows a convenient and efficient transport of the complete system by truck, train, ship or airplane.

2. CONCEPT OF THE MODIFIABLE UNDERWATER MOTHERSHIP (MUM) AND REGULATORY CHALLENGES

For the operation of offshore infrastructure, a series of permanent underwater works are necessary. Figure 1 illustrates some of the most important tasks. The production and transportation of fossil fuels require e.g. an intensive use of remotely operated vehicles (ROV), which are applied for underwater work or maintenance such as controlling of valves. The ROV is usually launched from a surface vessel and a permanent cable connection from the ROV to the surface vessel is required. Even if the actual control of the vehicle is mostly carried out by experts at onshore station (data transfer is realised via satellite communication), the operation of the surface vessel still requires a certain amount of crew. In addition, the weather and wave conditions must not be too bad, to ensure a safe operation not only for the crew but mainly also for launch and recovery of the ROV or other robots. In reality, operators of offshore service vessel expect a considerable number of days of downtime per year, due to inoperable weather conditions.

The new design of the unmanned underwater vehicle (UUV) presented in this paper aims to replace many of the required offshore service vessels so far. This procedure considerably reduces the influence of weather conditions and thus maximizes the number of operation days per year. Furthermore, it minimized the required amount of crew members. An offshore crew on board of a service vessel is not required any longer. Only a small team of vessel operators might be required, which also could be located on an onshore base. This enormously reduces the normally very meaningful issue of safety on board.

Figure 1: Seabed Infrastructure Works

To fulfil the requirements mentioned above, the new Modifiable Underwater Mothership (MUM) requires an appropriate level of displacement to carry all of the required payloads for acting as a mothership and for being able to replace offshore surface vessels. The demonstrator currently in production will have a length of approximately 26 m, a width of 6.9 m and a height of 3.5 m. Figure 2 shows an isometric view of the MUM demonstrator and table 2 lists its main properties.

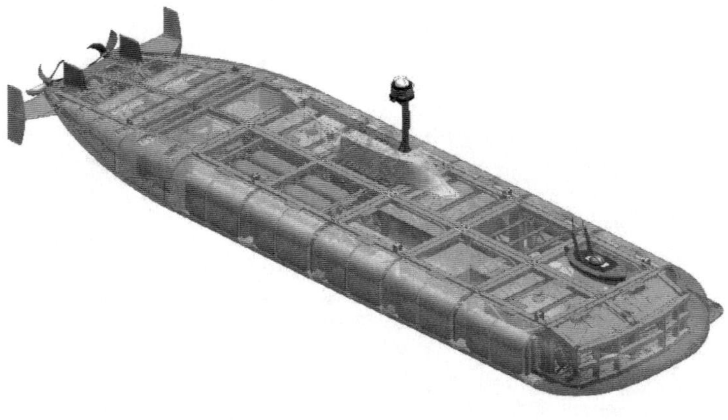

Figure 2: Modifiable Underwater Mothership (MUM)

Table 1. Specifications/ Characteristics of MUM

Classification	XLUUV
Dimensions (Demonstrator)	26 m length, 7 m beam, 3.5 m height
Energy system	Electric, Fuel Cell supported by battery
Communication	RF Com (LOS), (tethered) SATCOM buoy, UW Com (modem), communication buoys
Depth rating	150+ m (Demonstrator)
Maximum range	1400+ nm
Maximum speed	8+ knots
Manoeuvring capabilities	Dynamic positioning (in current up to 2 kn)
Payload capacity	Flexible / Mission adaptive

As visible in Figure 2 the MUM main structure is characterised by a dismountable steel frame. Disassembled, each components (also called module) can be transported as a 20 foot or 10 foot container and, thus, easily handled and transported to almost any location in the world. The second significant advantage of this modular system is that the modules are interchangeable and replaceable. A distinction is made between basic module and mission module. Basic module are mandatory for the vehicle operation, such as energy supply modules, trim and diving cell modules or propulsion engine module. Mission modules, on the other hand, can be customized by the MUM operator and very specifically adapted to the requirements of MUMs use case.

The left side of Figure 3 shows exemplary a basic module. This is a combined fuel cell and reactant module that ensures the supply of electrical energy. The fuel cell is mounted inside a pressure hull in front of the image. The back of the image shows the reactant storage. Oxygen and hydrogen are stored in high pressure canisters. The right side of Figure 3 shows an example of a mission module. This is the communication buoy module. The communication buoy carries typical communication devices, such as a satellite communication system, that's allows to transfer high density data from MUM to a command and control center or to remotely control an ROV or even the entire MUM.

Figure 3: MUM Modules

The demonstrator of MUM consist of the following basic module:

- Aft Ship
- Fuel Cell Module
- Reactant Module
- Battery Module
- Trim and Diving Module 1&2
- Mast Module
- Side Casing
- Fore Ship

And of the following mission module:

- Remote Operating Vehicle Module (ROV Module)
- Communication Buoy Module
- Acoustic Modem Module (five AUV that can be launched and recovered autonomously)
- Future Growth Module

The demonstrator as such does not reach reflect the full capability of the final product, but focusses on areas for research necessary from the point of view of a seasoned submarine builder. Enhancements of the demonstrator can easily be made by the inclusion of further modules

If the operator e.g. request for extending the range of the vehicle, he is able to add further Reactant Module or if he needs more electrical power he could add a second Fuel Cell Module. Furthermore, customized modules that fulfil individual requirements of the operator (e.g. a Drilling Rig Module) can be added to the vehicle as well. Adding further modules increase the total length of the vehicle. At the present time, a maximum length of 50 m is allowed. This restriction follows mainly from the requirement that MUM can be used as a seagoing vessel in surfaced operation and that the global loads, such as global bending or torsional moments, increase quadratically with the vehicle length. The demonstrator can operate up to sea state 3.

The main propulsion system of MUM is based on to contra-rotation propellers that are powered by two individual electrical motors. This design provides efficient transit capabilities in submerged and surfaced operation. Classical steering manoeuvres are handled by in total four side rudders and two diving planes. Furthermore, MUM is designed for dynamic positioning. Two Pump Jets (supplied by Schottel), one near the stern and one near the bow, allow accurate positioning in the horizontal plain in currents up to two knots.

The capability to deploy objects to the seabed is enabled by the trim and diving modules in cooperation with four vertical thrusters. Releasing an object may change the weight and buoyancy relation abruptly. This deviation can be corrected instantaneously by four vertical thrusters. In parallel the trim and diving system adapts the unbalance until an equilibrium is reached and the thrusters can be disengaged.

3. TECHNICAL DESCRIPTION OF MUM

3.1 MAIN PROPULSION AND ENERGY SUPPLY

As shown in Figure 2, the main propulsion is based on two contra-rotation propellers. This propellers work with low rotational speeds and thus, achieve excellent propulsion efficiency and minimal noise radiation. The hydrodynamic properties and propulsion characteristics were deeply investigated during the vehicle design, see e.g. Kösterke et al. (2024) and Greve et al. (2022). Each drive train is powered by an individual electrical motor with a maximum electrical power of approximately 70 kW.

The electrical energy is supplied by a combined fuel cell and battery system. A visualisation of the fuel cell (mounted inside a pressure hull) and the high pressure cylinders for storing oxygen and hydrogen as well as reactive water was already shown in Figure 3. The battery pressure hull and the integrated lithium ion battery is shown in Figure 4. Due to the modular system structure, MUM can be extended with further fuel cells and batteries to increase the maximum deliverable electrical energy, considering also scenarios in which one of the main tasks of Mum on station might be to supply energy to the systems in operation off board MuM.

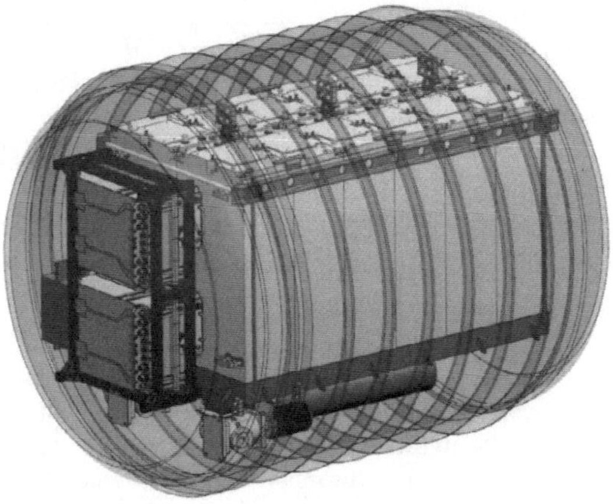

Figure 4: Battery Module

The electrical energy is supplied via two grids. Every module (i.e.: basic and mission module) is connected to both grids, thus, fulfilling highest requirements of redundancy.

3.2 MANOEUVRING SYSTEM

MUM is equipped with an extensive manoeuvring system that allows dynamic positioning in currents up to 2 kn. Figure 5 shows an isometric, side and top view of the stern with the steering system. Usual manoeuvers, such as zig-zag, turning circles manoeuvers or changes of the diving depth, are performed via four side rudders and two diving planes with winglets. The rudder pairs and diving planes can be controlled individually and thus fulfil redundancy requirements.

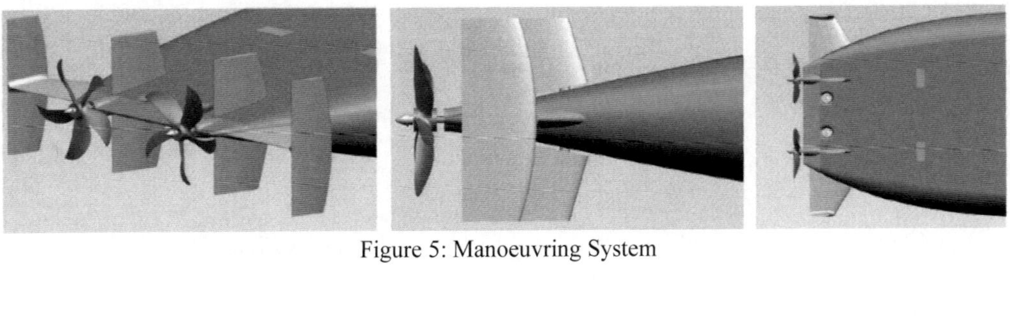

Figure 5: Manoeuvring System

For dynamic positioning, MUM is equipped with two horizontal thrusters (Pump Jets supplied by Schottel) that are able to rotate the thrust vector 360° in the horizontal plane. These thrusters allow dynamic positioning in currents up to 2 kn and enable manoeuvring in very limited space.

Four further thrusters are located vertically, two at the bow and two at the stern. The right side of Figure 5 shows exemplarily the stern vertical thrusters between the side rudders. This thrusters allow MUM launch and recover devices. Launching and recovering devices may change the hydrostatic properties and the equilibrium of mass and buoyancy related forces rapidly. The vertical thrusters are able to instantaneously rebalance the hydrostatic equilibrium of the vehicle by generating a vertical force, until the trim and diving modules adapted their filling level according to the new hydrostatic situation.

3.3 AUTONOMY AND NAVIGATION

The vehicle is equipped with a central controlling system (CIM) that is able to control all vehicle devices and systems. The data transfer is realized by two separate ring networks. A redundant data transfer is realised by connecting every module with both ring networks. Before starting an autonomies mission, the mission process will be predefined by use of a mission planning software and uploaded to CIM. A possible mission could be, for example:

- Leave the base and transit in surfaced mode to a predefined position
- After reaching the predefined position, dive to a diving depth of 50 meter and transit to the destination position
- Launch the communication buoy (more details in chapter 3.4) every 12 hours to correct the inertia navigation system with a gps position
- After reaching the destination, launch the communication buoy and activate remote control of the ROV for maintenance of subsea valve
- After finishing the job, proceed autonomously to the next destination

An autonomous operation requires a distinctive knowledge of the vehicle behaviour and its systems. The manoeuvring behaviour of the vehicle was analysed during development. Based on that knowledge, manoeuvring and control models are formulated to ensure a reasonable operation precision. Furthermore, this models can be improved continuously based on real operation data that will be measured during missions. The situation awareness of the environment and surrounding participants in maritime traffic is concluded based on a complex and interactive cooperation of sensors. An active collision avoidance sonar (mainly for submerged operations), a radar system and AIS comprises the most important devices. For accurate navigation a GPS, an inertia navigation and a seabed navigation system builds the core devices. The core of the

MUM is one of thyssenkrupp Marine Systems unmanned and autonomous vehicle fleet, which comprises submersible, semi-submersible and surface vessels. Each of these unmanned vehicles can be controlled by thyssenkrupp Marine Systems "Joint Autonomous Guidance and Operating System" (JAGOS). JAGOS is a software tool and serves as a monitoring system during autonomous operations or a control system during remotely control of the vehicles as well as their payloads. It generates a precise situation awareness manly based on radar, sonar, AIS and visual device data and manage collision avoidance in compliant with the rules for maritime traffic.

3.4 COMMUNICATION SYSTEM

The communication system provides the following lines of communication

- SATCOM
- Public Land Mobile Network (LTE)
- Public Land Mobile Network (WiFi/WLAN)
- AIS transceiver
- VHF Radio
- Accoustic underwarter communication

This systems are mounted at a hoistable mast at MUM and within a communication buoy, resulting in fully redundant communication capabilities. The data transfer rate of the SATCOM device (starlink) allow remote control of MUM, e.g. during ROV operations and the transfer of high density video or sensor data, thus, allowing MUM to perform complex operations under fully human monitoring and control.

4. RULES AND REGULATION

The operation of autonomous and unscrewed vessels is relatively new for maritime authorities as well as classification societies. Usually, submersible vessels of MUM-size are crewed and thus are subjected to rules and regulations that are not always reasonable for unscrewed vessels. The MUM project scope includes the development of specific rules and regulations for autonomous XLUUVs. This comprises rules for safe autonomous participation at maritime traffic as well as technical regulations, e.g. design rules for the wave-induced global structural loads of an XLUUV in surfaced operation. The complete development of all required rules and regulations for autonomous XLUUV operations will exceed the project timeframe and not be finish for the MUM sea trials. To ensure safe sea trials, a surface support vessel convoys the complete trial procedure. The sea trials will be performed in a delimit sea area.

For all of these aspects we are being supported and challenged by our developmental partners DNV, FKIE, DLR.

5. MISSION EXAMPLES

Based on the modular design philosophy and on the capability of carrying and operating heavy and energy consuming payloads, MUM can be applied for a wide range of application. As aforementioned, MUM is capable of replacing Offshore Supply Vessels fully for operations such as subsea inspections, maintenance or repairing. Figure 6 illustrates a typical ROV operation. MUM does not need to be transported offshore via a supply vessel. It is able to start its mission from the harbour and to transit autonomously to the sea area of operation. Once there, MUM launches the communication buoy, opens a satellite communication gateway to the command and control center and enables a remote control of the ROV and even of the complete MUM.

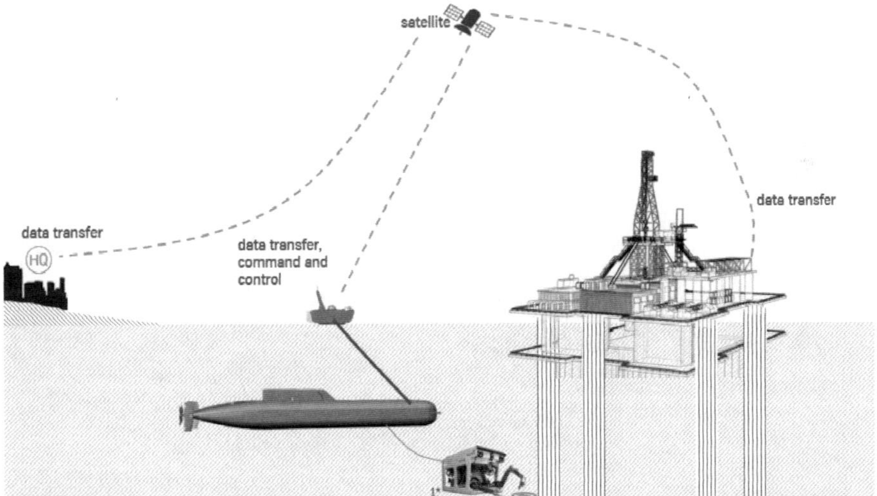

Figure 6: MUM mission as offshore underwater Mothership. (Source of the ROV: https://www.innova.no/e/products/rov-systems/sub-atlantic-comanche)

Since some years, critical infrastructure have increasingly become the focus of terrorist attacks. It is the responsibility of the operator of these infrastructures to secure their systems. MUM is the ideal vehicle to search for manipulations and thus could help significantly to protect critical infrastructures. Figure 7 shows possible protection mission. Equipped with a synthetic aperature sonar, MUM could autonomously scan critical infrastructures (such as gas pipelines) in regular intervals. An artificial intelligence could compare the current sonar data with the previous data that has been recorded during the last scan operation. When the system identifies undefined changes or manipulations, MUM could immediately launch the communication buoy and initiate a direct communication gateway to a command and control system, thus, minimum response times and quick countermeasures can be achieved. The ROV could be activated and applied to perform inspections immediately.

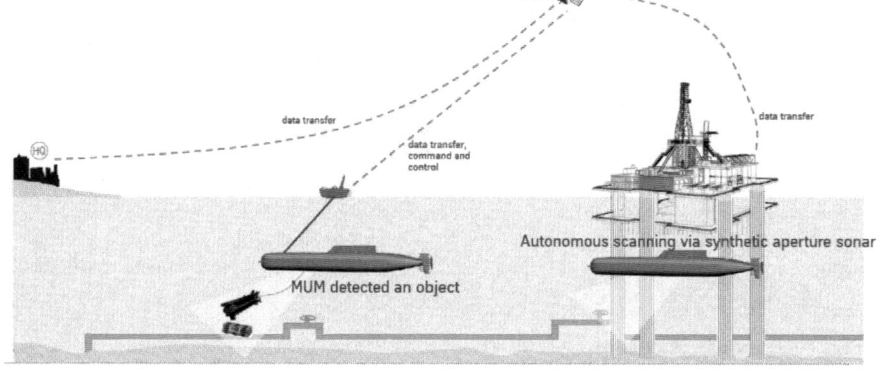

Figure 7: MuM mission as an autonomous inspection vehicle for critical infrastructures.

The aforementioned attack of critical infrastructures could be performed with underwater vehicles. To stop the attack at an early stage, MUM could be applied as a patrol vessel basically acting as an extended sensor. Due to MUMs main dimensions, it is capable to carry a sonar suit with a comparable performance to military submarines. Example given, it could carry a towed array and carry flank arrays or a conformal array to detect an approaching vehicle early. The petrol operation could be performed fully autonomous and whenever the artificial intelligence detects a targeting sonar signal, further measures could be initiated by launching the communication buoy. Figure 8 illustrated the petrol operation of MUM.

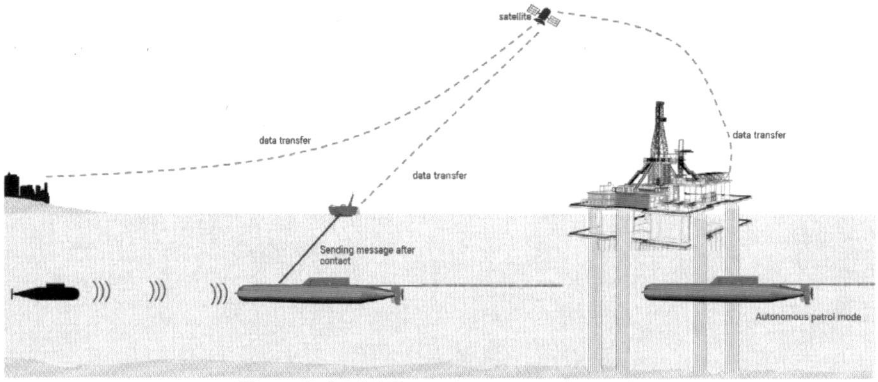

Figure 8: MuM Operation as patrol vessel.

6. SUMMARY

The main goal of the current project phase MUM2 is to develop, build and prove the concept with a full-scale demonstrator. Over-all, the demonstrator will have a length of approximately 25 meters. The initial research MUM1 already demonstrated the feasibility of a modular XLUUV kit from 2017 to 2020.

MUM2 started in spring 2021 and scheduled to end in early 2025. The project will transfer thyssenkrupp Marine Systems' technological leadership in non-nuclear submarines and know-how of small scale AUVs and UUVs to extra-large unmanned underwater vehicles and their respective systems.

7. ACKNOWLEDGMENTS

The research is supported by the Federal Ministry for Economic Affairs and Climate Action under Grand o3SX439.

8. REFERENCES

1. BMT Agross BV (2013), www.waveclimate.com, Marknesse, The Netherlands

2. DICK, R F., 'Predicting offshore workability for platform supplyvessels using IoT and machine learning', *Maritime Institute Willem Barrentsz, University of Applied Science, Terschelling, Netherlands*, 2023.

3. GREVE, M., RENTZOW, E., KUROWSKI, M., RITZ, S., GOLZ, M., BAYAZIT, N., and VIJAYASARATHI, L N., 'Design of the Propulsion System for the autonomous XLUUV MUM', *Proceedings of the ASME 2022 41st International Conference on Ocean, Offshore and Arctic Engineering OMAE2022, Hamburg, Germany*, 2022.

4. HOGBEN, N., DA CUNHA, N M., and OLIVER, G F., 'Global Wave Statistics', *British Maritime Technology Ltd., Feltham, Middlesex, UK*, 1986.

5. KÖSTERKE, M., VOß, J P., RIESNER, M., and GREVE, M., 'Propeller Design and Dynamic Hull-Propeller Interaction of an XLUUV', *8th International Symposium on Marine Propulsors, Berlin, Germany*, 2024.

6. MAURO, F., and NABERGOJ, R., 'A probabilistic approach for Dynamic Positioning capability and operabilitypredictions', *Ocean Engineering, Vol 262*, 2022.

7. SÖDING, H., 'Global Seaway Statistics', *Schriftenreihe Schiffbau, Report No. 610, Technical University Hamburg-Harburg, Hamburg*, 2001.

8. VEER, R., VAN'T AND GACHET, M., 'Dynamic Positioning – Early Design, Capability And Offsets, A Novel Approach', *30th International Conference on Ocean, Offshore and Arctic Engineering, OMAE2011-49354, Rotterdamm, Netherlands*, 2011.

AUTONOMOUS SWATH VESSELS FOR WIND FARM INSPECTIONS: A COMPARATIVE STUDY ON OPERATIONAL EFFICIENCY

S A Korkmaz, S R Turnock and **D Taunton,** Maritime Engineering, University of Southampton, UK
M Prince, Wolfson Unit MTIA, University of Southampton, UK
M Tinmouth and **N Tinmouth,** Acua Ocean, UK

SUMMARY

This study investigates the performance of autonomous SWATH vessels, both with and without foils, in offshore wind farm (OWF) operations near the Shetland Islands. Utilizing hybrid propulsion systems powered by Proton Exchange Membrane Fuel Cells (PEMFCs) and batteries, the research focuses on the vessels' operational availability during critical ROV deployment tasks. Through numerical simulations and a case study, the SWATH without foils consistently outperforms traditional monohull vessels in terms of stability and availability, particularly under challenging sea conditions. The findings highlight the advantages of SWATH vessels in maintaining consistent performance in harsh environments, crucial for the efficiency and safety of OWF maintenance. This study offers valuable insights into the operational benefits of adopting advanced SWATH vessel designs, demonstrating their potential to enhance the sustainability and effectiveness of offshore wind farm operations

1. INTRODUCTION

As the shift to renewable energy accelerates, Offshore Wind Farms (OWFs) have become crucial for generating clean energy and meeting global demands while reducing fossil fuel dependency. Positioned to harness strong offshore winds, OWFs play a significant role in the renewable energy mix. However, their success depends on regular inspection, repair, and maintenance (IRM) operations, which are vital to prevent failures and ensure continuous operation [1]. These operations are challenging due to harsh sea conditions, including high waves, strong currents, and unpredictable weather, which complicate maintenance tasks and pose significant health and safety risks [2]. As OWFs expand into deeper waters, the need for more reliable and efficient IRM methods becomes increasingly critical.

Despite advancements in inspection technologies and optimization strategies, existing research has largely focused on operational processes rather than evaluating the operational availability of vessels specifically designed for autonomous OWF inspections [3-5]. The comparative efficiency of vessel types, such as monohulls and SWATH vessels, particularly for autonomous operations like ROV deployment, remains underexplored.

This study addresses this gap by designing and analysing an autonomous SWATH vessel powered by a hybrid propulsion system that combines Proton Exchange Membrane Fuel Cells (PEMFCs) and batteries, using liquefied hydrogen. It focuses on the vessel's operational efficiency and seakeeping performance in challenging offshore environments, offering new insights into their potential for OWF IRM operations.

A case study compares the availability of SWATH and monohull vessels at a future OWF location near the Shetland Islands, advocating for a shift from traditional diesel-powered monohulls to renewable energy-driven autonomous maritime technologies.

The paper details the research approach, including vessel design, mathematical modelling, and a case study, and concludes by emphasizing the importance of SWATH vessels in OWF maintenance, with implications for future research and industry practices.

2. REVIEW OF ADVANCEMENTS IN OFFSHORE WIND FARM MAINTENANCE AND INSPECTION TECHNOLOGIES: A FOCUS ON VESSEL DESIGN AND SEAKEEPING PERFORMANCE

The rapid expansion of OWFs has necessitated advancements in IRM technologies to ensure the reliability and longevity of these critical energy infrastructures. As OWFs are increasingly deployed in more remote and challenging environments, the need for efficient, cost-effective, and reliable maintenance strategies has become paramount. This section reviews the latest research and technological developments related to the design and seakeeping performance of vessels used in OWF IRM operations. It highlights the evolution of maintenance strategies, the integration of autonomous systems, and innovations in vessel design, particularly focusing on the role of SWATH vessels. These studies provide a comprehensive understanding of the current state of OWF maintenance, and the challenges associated with vessel performance under varying sea conditions, setting the stage for further exploration into the operational efficiency and environmental impact of these vessels.

Various studies have addressed the challenges and advancements related to OWF inspection and maintenance. For example, one study [6] emphasize the importance of risk-based maintenance planning in ensuring the reliability and extending the operational lifespan of offshore wind turbine farms. Their study highlights that effective inspection and maintenance strategies are crucial in managing the life-cycle costs and mitigating the risks associated with structural failures in offshore wind farms. Another report [7] estimate that the operations and maintenance costs could constitute up to 29.8% of the total lifecycle costs for these floating installations, underscoring the economic significance of effective maintenance strategies. A cost-benefit evaluation [8] of remote inspection technologies for offshore wind farms, demonstrating that these systems can provide significant economic benefits by reducing the need for expensive and time-consuming access to turbines. Similarly, a cost-benefit analysis [9] of robotics-driven inspections for floating OWFs, focusing on the operational expenditure and efficiency gains from using a multi-robot platform. While the study highlights the economic advantages of robotic systems in these operations, it does not delve into the specific types of vessels that would be most effective for deploying these systems.

Optimization strategies have also been a key focus in recent research, aimed at improving the operational efficiency of OWF inspections. A recent study [10] proposed a comprehensive boat-assisted drone inspection scheme for OWFs, emphasising the importance of optimal energy consumption and efficient route planning for both the drone and the supporting boat. While the study demonstrates the effectiveness of using drones for wind farm inspections, it also underscores the need for reliable boat platforms to serve as movable control centres and battery swap stations, especially in challenging offshore environments. Another research effort [11] developed a trajectory planning algorithm based on the Ant Colony Optimization metaheuristic algorithm, specifically designed to optimise the number of Autonomous Surface Vehicles (ASVs) and their routes for efficient inspection of OWFs. The study demonstrates how this approach can minimise the total travel distance and ensure complete observation of offshore structures, highlighting its potential for improving Operations and Maintenance (O&M) actions. Additionally [12] proposed an optimization approach for OWF inspection routes using a combination of K-means clustering and Genetic Algorithm (GA). Their study focuses on minimizing inspection route lengths and operational costs for inspection vessels.

One of the significant areas of research has focused on the role of autonomous and unmanned systems in OWF operations. A 2022 study [13] highlighted the potential of collaborative unmanned vehicles, including Unmanned Aerial Vehicles (UAVs), Unmanned Surface Vehicles (USVs), and Unmanned Underwater Vehicles (UUVs), for IRM operations of OWFs, underscoring the importance of reliable and efficient platforms for such operations in challenging offshore environments. Another research effort [3] further explored the use of ASVs equipped with both emersed and immersed sensing capabilities for the multi-domain inspection of OWFs. Their study highlights the efficiency of using a single ASV to generate comprehensive 3D maps that integrate data from both above and below the water surface, thereby enhancing the overall inspection process.

Another research area was about the operational challenges associated with OWF maintenance have been extensively analysed. A 2024 review [14] conducted a comprehensive review of multi-robot coverage path planning for OWF inspections, emphasizing the challenges and potential of autonomous systems in this domain Another comprehensive review [15] provides a review of monitoring, operation, and maintenance practices for smart OWFs, emphasizing the integration of advanced technologies such as digital twins and predictive maintenance to enhance efficiency and reliability. An accessibility analysis [16] conducted a comprehensive accessibility analysis for OWFs in the North Sea, focusing on the operational challenges posed by met ocean conditions for maintenance activities. While the study provides insights into the accessibility challenges for different vessel types, including workboats and helicopters, it does not specifically address the performance of vessels or the environmental impacts of propulsion systems.

Recent studies have increasingly focused on enhancing the seakeeping performance and operational efficiency of SWATH vessels in offshore conditions. One study [5] investigated the use of electric and hybrid propulsion systems in SWATH vessels, highlighting significant reductions in energy consumption and greenhouse gas emissions. The research emphasized the hydrodynamic benefits of SWATH designs, such as reduced sway and heave, which are vital for stability in varying sea conditions. Advanced simulations demonstrated improved stability and environmental impact through hybrid energy storage systems combining batteries, hydrogen fuel cells, and diesel generators. Another study [17] assessed the impact of fin stabilizers on semi-SWATH vessels, finding a notable reduction in pitch angle, enhancing stability during operations. Research on Stabilizing Ordinary Fishing Nets (SOFNets) [18] showed their effectiveness in reducing heave and pitch, offering a cost-effective stabilization method for SWATH vessels. Further investigation [19] into anti-rolling devices revealed their potential to enhance stability and reduce rolling motion in harsh sea conditions, thereby improving operational reliability. Additionally, the Operability Robustness Index (ORI) [20], though initially focused on monohulls, was introduced to optimize seakeeping performance, with principles applicable to SWATH vessels. This research underscored the importance of balancing vessel size and hull design to enhance efficiency and reduce costs. Finally, an operability analysis [21] comparing SWATH vessels with monohulls for offshore wind turbine maintenance demonstrated SWATH vessels' superior operability in high sea states, essential for walk-to-work operations. However, it also noted higher fuel consumption due to the unique SWATH design and limited damping.

3. DESIGN SPECIFICATIONS OF SWATH AND CONVENTIONAL MONOHULL VESSEL

This section compares the design specifications of the autonomous Hybrid Unmanned Surface Vessel (H-USV) and a traditional crewed monohull vessel used in offshore operations. The H-USV, shown in Figure 1, is a SWATH vessel designed for autonomous missions, minimizing wave-induced motions crucial for safe ROV deployment. It uses a hybrid propulsion system powered by PEMFCs and batteries, with liquefied hydrogen as fuel, optimized for a speed of 3.5 knots.

Figure 1: Structural Overview of the Constructed H-USV.

The conventional monohull vessel, larger and crewed, operates at a higher speed of 10 knots, powered by two 331 kW diesel engines. While this design supports extended missions with a crew of 20 for up to four weeks, its susceptibility to wave-induced motions can impact operations, particularly during ROV deployment [22-23].

In contrast, the H-USV is engineered for long-duration, autonomous missions, with a 40 kW PEMFC and two 120 kWh lithium-ion batteries, enabling up to 40 days of operation without refuelling or crew support. The monohull excels in speed and capacity, whereas the H-USV offers superior endurance and autonomy, catering to different offshore needs. Key dimensions of both vessels are summarized in Table 1. Various remote locations on each vessel were analysed for vertical motions, velocities, and accelerations—key factors in determining the operational limits. These locations could be found in the report [23].

Table 1. Dimensions of monohull boat and SWATH [23].

Parameters	Monohull	SWATH	Units
Overall Length	38.97	13	m
WL Length	36.49	12.25	m
Beam max extents on WL	8.9	9.25	m
Draft Amidships	2	1.6	m
Wetted Area	317.6	-	m^2
Displacement	257.7	24.065	T
LCG	17.89	6.78	From zero pt. (+ve fwd) m
VCG	2	2.29	From baseline m
Kxx	3.56	3.6	m
Kyy	9.0	3.51	m
Kzz	9.0	3.51	m
Cut Water Plane Area	281.57	5.01	m^2

4. SEAKEEPING METHODOLOGY

This section outlines the seakeeping methodology employed to evaluate the hydrodynamic performance of both the H-USV and the conventional monohull vessel. The assessment was conducted using ANSYS AQWA [24], a specialized software designed to analyse vessel behaviour in response to wave interactions. Ansys AQWA employs three-dimensional radiation/diffraction theory in the frequency domain to simulate linearised hydrodynamic fluid wave loading on floating bodies, making it particularly well-suited for simulating the complex interactions between vessels and the marine environment. This software provides detailed insights into factors such as heave, pitch, and roll motions, which are critical for assessing vessel stability.

4.1 MESH GENERATION AND VESSEL MODELLING

Ansys AQWA utilizes a surface mesh composed primarily of quadrilateral panels to model the vessels. For the monohull vessel, characterized by its sharp chine, many panels were required to accurately represent its complex geometry (Figure 2 (a)). In contrast, the SWATH vessel (Figure 2 (b)) required careful consideration in the panel distribution, particularly for the hull shapes and displacement calculations. Notably, the orthogonal surfaces of the SWATH's cross deck were excluded from the model, as only the geometry below the waterline is essential for the analysis. A limitation of the AQWA meshing algorithm is that it does not account for symmetry, resulting in slight discrepancies between the port and starboard sides, which introduced minor asymmetries in the response results.

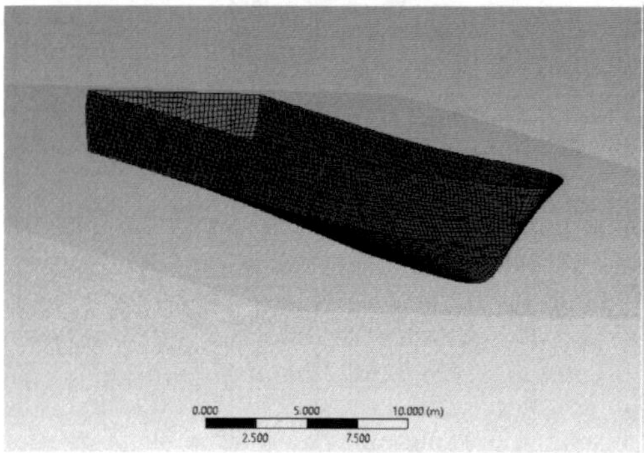

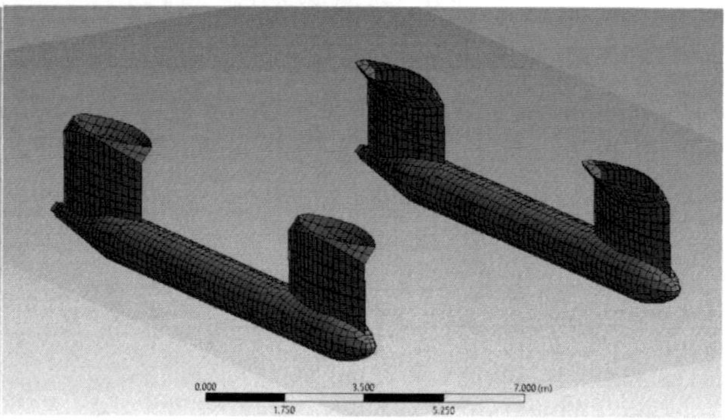

(a)

(b)

Figure 2: Monohull (a) and SWATH (b) mesh in Ansys AQWA [23].

4.2 HYDRODYNAMIC SIMULATION AND ASSUMPTIONS

The seakeeping performance of both vessels was simulated for various forward speeds, approximated for conditions where the Froude number (Fr) is less than 0.3, consistent with limitations outlined in [24]. The monohull vessel's performance was modelled with a focus on its response to different wave conditions without additional damping corrections, reflecting the typical behaviour of crewed vessels.

For the H-USV, additional enhancements were necessary to accurately predict performance. The study incorporated additional numerical models to account for viscous damping and the lift generated by the vessel's foils, which are critical factors for the SWATH design. The method used for including these effects is based on the approach developed by [25] and more recently applied by [26] The approach used is illustrated in the flow chart in Figure 3. This iterative method updates the hydrodynamic coefficients to incorporate the effects of viscous damping and foil-generated lift, significantly reducing the amplitude of vessel motions in response to waves. These modifications were essential for providing a more realistic assessment of the SWATH vessel's stability and operational availability in various sea states.

By integrating these modelling approaches, the study achieved a detailed and accurate simulation of the vessels' hydrodynamic behaviour under various sea states. This provided essential insights into the operational reliability and stability of both vessel types, particularly in the challenging conditions of the North Sea

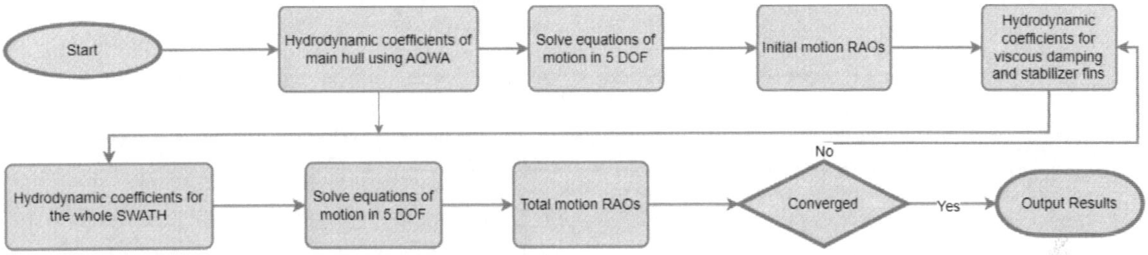

Figure 3: Flowchart showing the modelling process for SWATH [23].

4.3 SEA STATES AND VESSEL SPEEDS

The study considered a range of sea states typical of the North Sea, characterized by significant wave heights (H_s) of 1.0m, 2.0m, 3.0m, and 4.0m. These sea states were selected based on data from the NOAA Wavewatch III hindcast model, which is hourly based one year data, specifically focusing on a prospective OWF location near Shetland Islands, which represents the challenging conditions vessels frequently encounter in this region. The JONSWAP spectrum was used to model wave conditions, as it accurately reflects the steep and short-period waves that are common in the North Sea.

Tank test experiments indicated that the peak in the pitch Response Amplitude Operator (RAO) occurred at a modal period of approximately 6.5 seconds. Consequently, the final selection of wave spectra included modal periods of 4.0s, 5.0s, 6.5s, 8.0s, and 9.0s. These modal periods are critical for assessing the vessels' performance under realistic maritime conditions. Since the model is linear, the response of the vessel in any wave height can be determined by scaling the responses in 1m significant wave height by the desired wave height, though this approach does not apply to the Motion Sickness Index (MSI) and Motion-Induced Interruptions (MII) results for the monohull vessel.

Vessel speeds were set at 0 knots (stationary), 3.5 knots, and 5 knots for the SWATH vessel, and 0 knots, 4 knots, and 10 knots for the monohull vessel. These speeds were chosen to reflect typical operational scenarios during ROV deployment and recovery. The primary focus of the study was on the vessel's performance at 0 knots, particularly during stationary ROV launching and recovery operations.

4.4 LIMITING CRITERIA AND AVAILABILITY INDEX

The limiting criteria for both the SWATH and monohull vessels were derived from established literature and equipment specifications [27].

The limiting criteria for the SWATH and monohull are given in Table 2 (a) and 2 (b), respectively. For the monohull vessel, which is a crewed vessel, the launch and recovery of the ROV are manual operations. Therefore, the limiting criteria are primarily based on human performance factors such as the MSI and MII due to sliding on deck or tipping over.

For both vessels, however, there are also limiting forces associated with the ROV launch and recovery process. DNV recommended practice [28] outlines calculations to determine the forces on an ROV, but there are no limits specified.

Table 2. Limiting criteria for SWATH (a) and monohull (b) vessels [23].

Criteria	Limi
Wet Deck Slamming	Freeboard = 1.15 m Critical Velocity = 1.05 m/s
Bow Keel Slamming	Draught = 1.6 m Critical Velocity = 1.05 m/s
Foil Emergence	Draught = 1.0 m
Propeller Emergence	Draught = 1.04 m
Fuel Cell Vertical Acceleration	9.81 m/s^2
Antenna Resultant Acceleration	9.81 m/s^2
Mast Top Resultant Acceleration	9.81 m/s^2
ROV Bottom Slamming	Freeboard = 0.7 m Critical Velocity = 1.05 m/s
ROV winch relative velocity	Critical Velocity = 1.05 m/s

(a)

Criteria	Limit
Pitch	3 degrees
Roll	8 degrees
Water on Bow Deck	Freeboard = 2.5 m
Bow Keel Slamming	Draught = 2.0 m Critical Velocity = 1.755 m/s
Propeller emergence	Draught = 2 m
ROV winch relative velocity	Critical Velocity = 1.755 m/s
Motion Sickness Index in % of crew in 120 minutes	
Motion-Induced Interruptions - Sliding	Friction = μg = 0.7*9.81
Motion-Induced Interruptions – Fore/Aft Tipping	Stance = $(L/h)*g$ = 0.17*9.81
Motion-Induced Interruptions – Side/Side Tipping	Stance = $(L/h)*g$ = 0.25*9.81

(b)

This study particularly focused on the slamming force experienced by the ROV as it is lowered to the sea surface. The slamming force F_{Slam} is calculated using the equation:

$$F_{Slam} = 1/2\, \rho C_s A_p V_s^2$$

where:
C_s is the slamming coefficient,
A_p is the horizontal projected area of the ROV $[m^2]$
V_s is the slamming impact velocity $[m/s]$, given by:

$$V_s = V_c + \sqrt{V_{ct}^2 + V_w^2}$$

where:
V_c is the hook lowering velocity,
V_{ct} is the characteristic single amplitude vertical velocity of the crane tip, and
V_w is the characteristic vertical water particle velocity, all measured in meters per second $[m/s]$.

The slamming force for a given ROV is dependent on the relative velocity of the crane winch relative to the sea surface. Without a limiting value for the force or velocity, the critical slamming velocity used for the hull and wet deck slamming is used.

The limiting criteria are used to calculate the probability of exceeding a critical value. The probability of exceedance is calculated using the Rayleigh probability distribution:

$$P_{exceed}(z > z_{crit}) = exp\left(-\frac{z_{crit}^2}{2m_{0z}}\right)$$

where:
z_{crit} is the limiting criteria,
m_{0z} is the variance of the response spectrum.

The availability index for each vessel was then calculated by determining the maximum probability of exceeding any of the limiting criteria under different sea states and speeds. This index is expressed as:

$$Availability\ Index = 100\% - \max\left(P_{exceed}\right)$$

This index represents the percentage of time that a vessel can safely operate without exceeding the limiting criteria. A higher availability index signifies greater operational reliability, which is particularly critical for the continuous, unmanned operations envisioned for autonomous SWATH vessels in offshore environments. Although this index provides an easy method of comparing the performance of the two vessels in the same conditions, it does not identify which specific criteria are the limiting factor.

By combining these criteria, the study was able to provide a clear comparison of the operational reliability of the SWATH and monohull vessels, offering valuable insights into their suitability for ROV deployment in the harsh conditions of the North Sea.

5. CASE STUDY: FUTURE OFFSHORE WIND FARM NEAR SHETLAND ISLAND

This case study aims to compare the operational performance of an autonomous SWATH vessel and a monohull conventional vessel at a prospective offshore wind farm (OWF) location near the Shetland Islands coordinates at 60.324°N 0.511°W. The location was chosen for its relevance to real-world conditions, characterized by frequent high sea states and strong winds typical of the North Sea. The study compared the operational availability of the SWATH vessel and a conventional monohull vessel, focusing on the availability of ROV launches and recoveries at the specified OWF location over one year. For this model, hourly data generated by the NOAA Wavewatch III hindcast model were utilised. Given that the study compares the ROV launching and recovery capabilities of the SWATH vessel with those of a conventional monohull vessel, the significant wave heights, wave periods, and wave directions in the selected region were analysed. The results of this analysis are presented in Figure 4 (a). Similarly, an analysis of wave directions in the selected region was conducted and is presented in Figure 4 (b). This analysis, displayed in polar coordinates, provides a comprehensive overview of the prevailing wave directions, which is crucial for understanding the operational environment and its impact on the vessel's seakeeping performance.

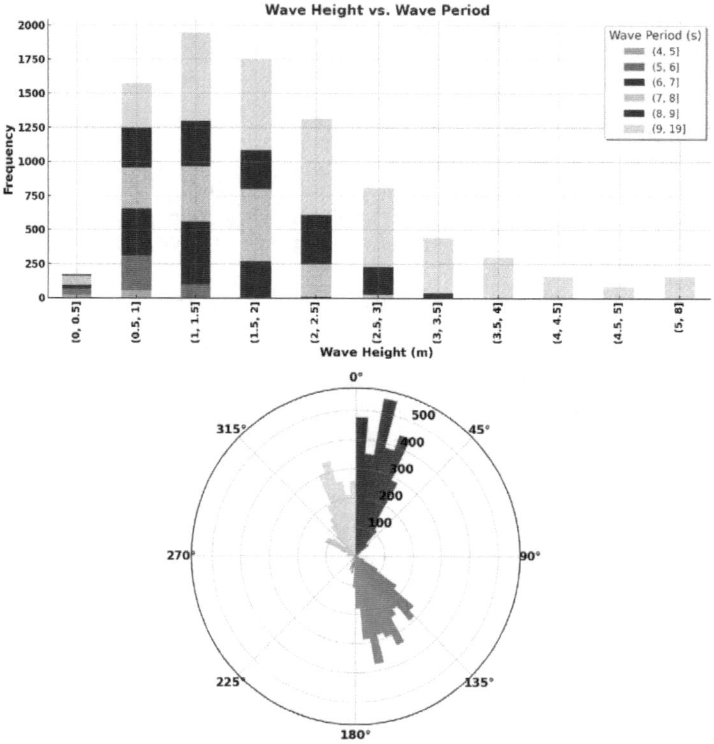

Figure 4: (a) Frequency of wave period and height combinations (b) Yearly wave directions with 5-degree bins.

6. RESULTS

6.1 OVERVIEW OF VESSEL AVAILABILITY PERFORMANCE

A comprehensive statistical analysis was conducted utilising ANSYS AQWA availability results for the vessels—Monohull, SWATH without foils, and SWATH with foils. The results offer insights into the overall availability of each vessel under varying conditions, as summarized in Table 3.

Table 3. Summary Statistics of Vessel Availability Under Varied Conditions.

Statistic	Monohull	SWATH without Foils	SWATH with Foils
Mean	74.25%	**86.58%**	84.83%
Standard Deviation	29.69%	**17.58%**	19.42%
Minimum	0.00%	**19.22%**	16.34%
25th Percentile	62.40%	**78.94%**	77.72%
Median (50th Percentile)	86.02%	**94.66%**	93.36%
75th Percentile	96.88%	**99.76%**	99.69%
Maximum	100.00%	**100.00%**	100.00%

The analysis shows that the SWATH without foils demonstrates the highest mean availability (86.58%) and the lowest variability (17.58%), suggesting that it performs more consistently across various conditions. The SWATH with foils, with a slightly lower mean availability of 84.83% and a higher standard deviation of 19.42%, indicates that the inclusion of foils introduces some variability, though the vessel still maintains high availability. In contrast, the Monohull, with the lowest mean availability (74.25%) and the highest standard deviation (29.69%), exhibits significant challenges in sustaining high availability, particularly under demanding conditions.

The SWATH vessels maintain higher minimum availability values compared to the Monohull, underscoring their better performance under challenging conditions. The median availability follows a similar pattern, with the SWATH without foils leading, followed closely by the SWATH with foils, and the Monohull lagging. The findings highlight the SWATH without foils as the most reliable vessel, particularly in operations requiring consistent performance.

In the analysis provided in Figure 5, which presents a symmetrical polar representation of vessel availability, the mean availability values are calculated across all speeds, wave heights, and wave periods for each heading. This comprehensive averaging offers a holistic view of how each vessel type performs directionally under varied conditions. We observe that the SWATH vessels, particularly the SWATH without Foils, exhibit more consistent and stable performance across all headings compared to the Monohull. This figure visually underscores the SWATH vessels' superior ability to maintain higher availability, regardless of the heading, with the SWATH without Foils excelling. This suggests that SWATH vessels are better suited for maintaining operational effectiveness across a broad range of conditions, offering a significant advantage in complex and variable maritime settings.

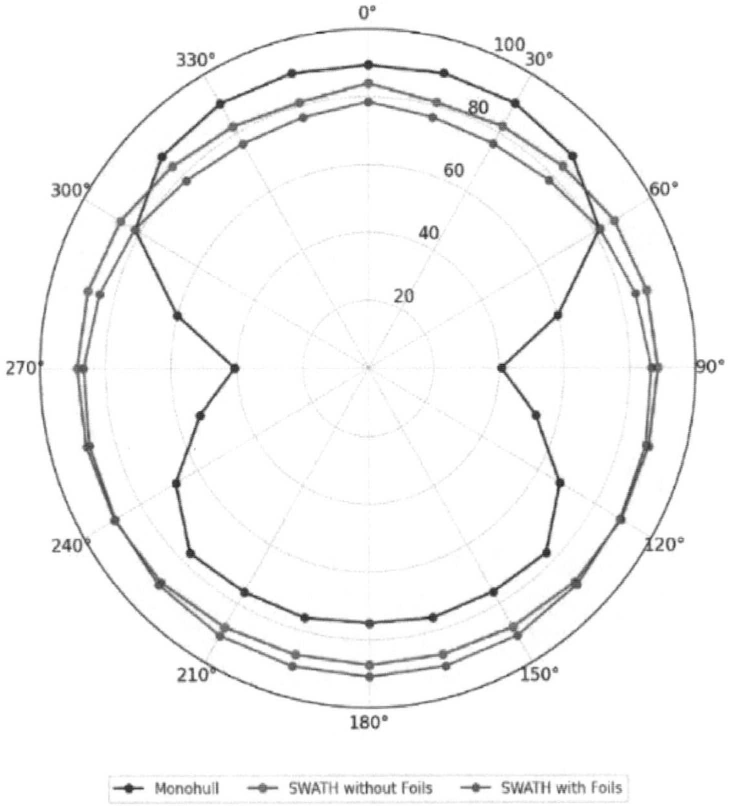

Figure 5: Symmetrical Polar Representation of Vessel Availability Across Varying Heading Angles.

Figure 6 provides a heatmap visualization of vessel availability, segmented by wave height and wave period across different speeds for each vessel type: SWATH with Foils, SWATH without Foils, and the Monohull. This figure illustrates the complex interaction between sea conditions and vessel speed on overall availability. The SWATH vessels, particularly without foils, maintain higher availability under varied sea states, as indicated by the cooler colours (blues) dominating the heatmaps. The Monohull, however, shows more red areas, indicating lower availability, especially under higher wave heights and longer wave periods. This visual comparison highlights the operational limitations of the Monohull in harsher sea conditions, further supporting the statistical analysis in Tables 4 and 5.

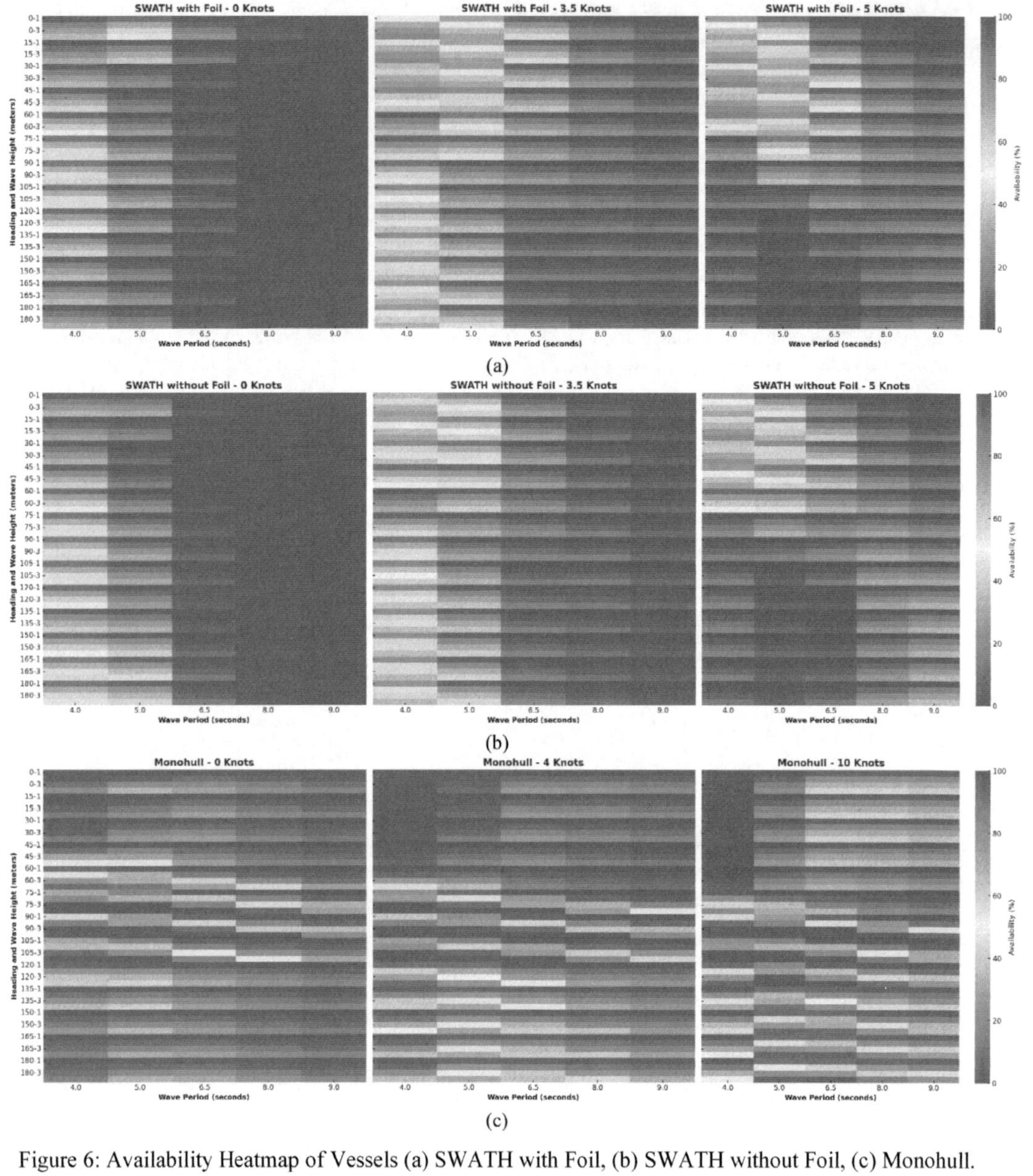

Figure 6: Availability Heatmap of Vessels (a) SWATH with Foil, (b) SWATH without Foil, (c) Monohull.

6.2 IMPACT OF SEA CONDITIONS ON VESSEL AVAILABILITY

The analysis of vessel performance across varying sea states—considering both wave height and wave period—reveals significant differences among the vessels. These conditions, collectively referred to as the "sea state," are critical in understanding how each vessel responds to the dynamic maritime environment. Table 4 and Table 5 summarize the availability statistics of the vessels under different wave heights and wave periods, respectively.

The combined analysis reveals that the SWATH vessels generally outperform the Monohull across varying sea states. Under lower sea states, characterized by a wave height of 1 meter or a shorter wave period of 4 seconds, the Monohull and SWATH vessels perform comparably in terms of mean availability. However, the SWATH without Foils leads slightly, especially in maintaining low variability.

Table 4. Impact of Wave Height on Vessel Availability Statistics.

Vessel Type	Wave Height (m)	Mean Availability (%)	Median Availability (%)	Standard Deviation (%)
SWATH with Foils	1	95.98	99.82	9.20
	2	88.44	95.76	16.44
	3	80.85	87.85	20.09
	4	74.05	79.42	22.02
SWATH without Foils	1	97.11	99.90	6.59
	2	90.09	96.85	13.83
	3	82.83	90.03	18.19
	4	76.30	82.26	20.78
Monohull	1	96.26	98.82	7.33
	2	83.18	91.43	20.35
	3	66.16	77.17	29.17
	4	51.39	60.06	32.62

Table 5. Availability Statistics by Wave Period.

Wave Period (sec)	SWATH with Foils Mean Avail. (%)	SWATH with Foils Median Avail. (%)	SWATH with Foils Std Dev (%)	SWATH without Foils Mean Avail. (%)	SWATH without Foils Median Avail. (%)	SWATH without Foils Std Dev (%)	Monohull Mean Avail. (%)	Monohull Median Avail. (%)	Monohull Std Dev (%)
4.0	68.27	70.44	24.58	69.55	71.60	22.60	70.86	89.28	34.78
5.0	76.72	81.20	21.44	81.09	85.12	17.54	65.92	79.82	33.80
6.5	89.56	95.05	12.73	92.77	97.24	9.65	70.67	82.13	29.87
8.0	94.04	98.87	7.95	94.29	99.08	8.40	79.12	87.13	24.35
9.0	95.55	99.64	6.39	95.23	99.70	8.69	84.66	91.97	19.18

As wave height increases or wave periods lengthen, the Monohull's performance deteriorates rapidly, particularly under the harsh conditions of 3 to 4 meters wave heights or wave periods extending beyond 6.5 seconds. The Monohull's availability drops to as low as 51.39% at a 4-meter wave height, with a significant rise in variability (32.62%), indicating increased unpredictability. Similarly, at a 9-second wave period, the Monohull's availability stabilizes at 84.66%, but it continues to show higher variability compared to the SWATH vessels.

The SWATH without Foils consistently shows the highest mean availability and the lowest variability across all wave heights and periods, making it the most stable and reliable vessel under varying sea states. The SWATH with Foils, while still outperforming the Monohull, demonstrates slightly lower availability and higher variability, likely due to the dynamic interactions between the foils and waves.

6.3 INFLUENCE OF VESSEL SPEED AND HEADING ON AVAILABILITY

This section examines the influence of vessel speed and heading on the availability of the two vessels—Monohull, SWATH with Foils, and SWATH without Foils. Both factors, vessel speed and heading, are critical determinants of operational effectiveness, as they directly impact the stability and reliability of the vessels under various maritime conditions.

Table 6 provide insights into how vessel speed affects availability across three aligned speed ranges: 0 - 0 knots, 4.0 - 3.5 knots, and 10 - 5 knots. At 0 - 0 knots, the analysis reveals that the SWATH vessels maintain higher mean availability compared to the Monohull. The SWATH with Foils achieves a mean availability of 91.65%, while the SWATH without Foils is slightly lower at 91.30%. In contrast, the Monohull lags with a mean availability of 78.25%. This significant difference underscores the superior design of SWATH vessels in maintaining stability and availability, particularly when stationary. Furthermore, the standard deviation for the SWATH vessels is significantly lower (13.32% for SWATH with Foils and 14.38% for SWATH without Foils) compared to the Monohull (27.80%), indicating more consistent performance for the SWATH designs.

Table 6. Vessel Availability Metrics by Speed.

Vessel Type	Vessel Speed (knot)	Mean Availability (%)	Median Availability (%)	Standard Deviation (%)
SWATH with Foils	0	91.65	99.21	13.32
	3.5	79.23	86.55	21.63
	5	83.60	91.78	20.23
SWATH without Foils	0	91.30	99.64	14.38
	3.5	83.88	92.38	19.12
	5	84.58	91.28	17.99
Monohull	0	78.25	89.01	27.80
	4	76.38	87.83	28.73
	10	68.12	79.36	31.55

As vessel speed increases to the 4.0 - 3.5 knots range, the Monohull's availability drops to 76.38%, while the SWATH vessels continue to outperform it, with 83.88% for SWATH without Foils and 79.23% for SWATH with Foils. The slight dip for the SWATH with Foils may be due to the complexity introduced by the foils at moderate speeds. The Monohull also shows the highest performance variability at 28.73%, compared to 19.12% and 21.63% for SWATH without and with Foils, respectively.

At higher speeds (10 - 5 knots), the disparity becomes more pronounced, with the Monohull's availability dropping significantly to 68.12%, much lower than the SWATH vessels (84.58% without Foils, 83.60% with Foils). The Monohull's variability remains the highest at 31.55%, while the SWATH vessels maintain more consistent performance.

Regarding heading, the SWATH vessels outperform the Monohull across all heading clusters. In the 0-15 degrees cluster, SWATH without Foils achieves the highest mean availability at 83.99%, with SWATH with Foils slightly lower at 78.55%. Although the Monohull shows a higher mean availability of 89.38% in this cluster, its stability is less consistent, with a lower standard deviation of 12.19%.

Table 7. Statistics of Vessel Availability across Different Heading Clusters.

Heading Cluster	SWATH with Foils Mean Avail. (%)	SWATH with Foils Median Avail. (%)	SWATH with Foils Std Dev (%)	SWATH without Foils Mean Avail. (%)	SWATH without Foils Median Avail. (%)	SWATH without Foils Std Dev (%)	Monohull Mean Avail. (%)	Monohull Median Avail. (%)	Monohull Std Dev (%)
0-15	78.55	88.79	24.51	83.99	95.51	21.50	89.38	94.66	12.19
15-30	76.60	86.27	25.67	81.07	92.16	23.88	89.95	95.33	11.72
30-60	77.32	86.12	24.52	83.22	92.71	20.51	89.27	93.98	12.03
60-90	83.29	88.55	17.60	87.61	93.40	14.06	70.91	84.97	32.60
90-120	87.30	92.28	14.47	88.79	93.86	14.54	46.85	47.97	38.12
120-150	89.57	95.66	14.38	89.23	96.71	14.93	72.28	81.22	26.94

150-180	90.87	97.28	14.48	87.69	95.68	16.49	76.07	86.06	25.24

As the heading shifts to 60-90 degrees, the Monohull's performance declines sharply, with a mean availability of 70.91% and a higher standard deviation of 32.60%, indicating increased variability and reduced predictability. In contrast, SWATH vessels maintain higher and more consistent availability, with the SWATH without Foils performing best at 87.61%. This trend is consistent across all headings, with the Monohull showing greater variability, particularly at larger angles. For example, at 150-180 degrees, the Monohull stabilizes somewhat with a mean availability of 76.07% but still lags behind SWATH vessels, which maintain superior performance at 90.87% (with Foils) and 87.69% (without Foils).

Overall, the combined analysis of speed and heading shows that SWATH vessels consistently outperform the Monohull, especially at higher speeds and challenging headings. Their lower variability suggests more predictable and reliable performance, making SWATH vessels better suited for operations requiring consistent availability. The Monohull, however, is less reliable under such conditions, impacting operational efficiency, safety, and cost.

6.4 RESULTS OF CASE STUDY

In this section, the comparative performance of the Monohull and SWATH without Foils (WoF) vessels is analyzed based on the availability scores derived from hourly data over a one-year period in the selected OWF region. The analysis focuses on how the availability of each vessel type aligns with the specific operational conditions experienced hourly throughout the year. It is important to note that these analyses were conducted with the vessels operating at 0 knots, meaning they were stationary during the data collection period.

The data reveals a clear distinction in performance between the Monohull and the SWATH WoF, with the latter demonstrating a significantly superior availability across all observed hours. As shown in Table 8, the average availability for the SWATH WoF is an impressive 99.85%, with minimal variability (standard deviation of 0.69%), indicating highly consistent performance. The SWATH WoF consistently maintains close to 100% availability, showcasing its robust design and capability to handle a wide range of maritime conditions with minimal disruption.

In contrast, the Monohull vessel exhibits an average availability of 96.23%, with a higher standard deviation of 4.24%. This increased variability underscores the Monohull's susceptibility to more significant fluctuations in performance, likely due to its less stable design when compared to the SWATH WoF. The Monohull's availability ranges from a low of 56.91% to a high of 99.79%, indicating that while it can perform adequately under certain conditions, it is prone to considerable performance drops under more challenging sea states in North Sea.

Table 8. Availability Analysis

Statistic	Statistic	Monohull Availability	SWATH WoF Availability
count	8737.0	8737.00	8737.00
mean	4369.0	96.23	99.85
std	2522.30	4.24	0.69
min	1.0	56.91	90.39
25%	2185.0	93.50	99.99
50%	4369.0	97.65	100.00
75%	6553.0	99.41	100.00
max	8737.0	99.79	100.00

Furthermore, the threshold analysis presented in Table 9 highlights the operational reliability of both vessels under varying conditions. For instance, the SWATH WoF consistently operates above a 95% availability threshold 99.74% of the time, whereas the Monohull only exceeds this threshold 71.36% of the time. This stark contrast in performance is even more evident when considering lower thresholds; the SWATH WoF remains operational above a 90% availability threshold 100% of the time, while the Monohull drops to 90.27%.

Table 9. Threshold Analysis

Threshold (%)	Monohull Above Threshold (%)	SWATH WoF Above Threshold (%)
70	99.86	100.00
75	99.74	100.00
80	99.65	100.00
85	99.24	100.00
90	90.27	100.00
95	71.36	99.74

These findings emphasize the SWATH WoF's clear superiority in terms of operational availability, making it a more reliable choice for maritime operations where high availability is critical. The Monohull, despite being a viable option, demonstrates limitations that could impact operational efficiency and safety, especially in less favourable sea conditions. This data-driven assessment underscores the importance of vessel selection in maritime operations and the advantages of choosing a vessel with higher and more consistent availability, such as the SWATH WoF.

This analysis could serve as a crucial input for decision-makers when planning operations in regions where sea conditions can be unpredictable, and high vessel availability is essential for mission success. The ability to maintain consistently high availability, as evidenced by the SWATH WoF, is a critical factor in ensuring operational effectiveness and reliability in challenging maritime environments.

7. CONCLUSION

In conclusion, this study has comprehensively analysed the operational availability of Monohull and SWATH vessels, both with and without Foils, focusing on their performance under various sea states and operational conditions within a prospective offshore wind farm (OWF) location near the Shetland Islands. The research included extensive hydrodynamic analyses at 0 knots, 3.5 knots, and 5 knots for the SWATH vessels and at 0 knots, 4 knots, and 10 knots for the Monohull vessel to determine their availability across a range of operational scenarios. These analyses were conducted with respect to defined wave heights, wave periods, and wave directions to simulate realistic maritime conditions.

Although these comprehensive analyses were performed, the case study portion of this research specifically focused on comparing the vessels' performance at 0 knots, which is critical for ROV launching and recovery operations. The findings reveal a distinct advantage of the SWATH without Foils (WoF) vessel over the traditional Monohull in terms of operational availability and consistency. With an impressive average availability of 99.85%, coupled with minimal variability, the SWATH WoF demonstrates its superior design and capability to maintain high operational efficiency across varying maritime conditions. In contrast, the Monohull vessel, with an average availability of 96.23% and higher variability, shows significant performance fluctuations, particularly under harsher sea states.

In addition to the case study findings, the general availability analyses across different wave heights, wave periods, and wave directions further emphasize the SWATH vessels' superior performance. The SWATH WoF consistently outperforms the Monohull vessel across all sea conditions tested, highlighting its robustness and reliability in challenging maritime environments.

These results suggest that the SWATH WoF is better suited for operations in environments where consistent high availability is essential. The implications of this study are significant for decision-makers in maritime operations, as they underscore the importance of selecting vessels that can reliably perform under demanding conditions, thus ensuring the success and safety of offshore wind farm maintenance and inspection activities.

However, the study is not without its limitations. The analysis focused primarily on stationary operations, and future research should explore the performance of these vessels under varying speeds and more dynamic conditions. Additionally, further studies could investigate the long-term operational costs and environmental impact of these vessel designs, particularly under different operational scenarios.

In summary, this research contributes to the growing body of knowledge on vessel performance in offshore environments and provides valuable insights that could guide future developments in maritime technology and vessel design for OWFs. The comprehensive analysis of both SWATH and Monohull vessels under realistic sea conditions offers critical information that can help optimize vessel selection and deployment strategies for OWF operations.

8. ACKNOWLEDGEMENTS

This research has been supported by the Hydrogen Innovation - Future Infrastructure & Vessel Evaluation and Demonstration (HI-FIVED) project (Project Number 10057707), funded under the UK Clean Maritime Demonstration Competition Round 3 (CMDC3), and the Green Shipping Corridors: Feasibility of the world's first hydrogen-powered North Sea crossing project (Project Number 10039998), funded under Round 2 (CMDC2) of the same competition, Innovate UK.

9. REFERENCES

1. GARCÍA MÁRQUEZ, M O., TOBIAS, A M., PINAR PÉREZ, J M., and PAPAELIAS, M., 'Condition monitoring of wind turbines: Techniques and methods', *Renewable Energy, vol. 46*, pp. 169-178, 2012.

2. BRINK, T., MADSEN, S O., and LUTZ, S., 'Perspectives on How Operation & Maintenance (O&M) Innovations Contribute to the Reduction of Levelized Cost of Energy (LCOE) in Offshore Wind Parks', *Danish Wind Industry Association*, 2016.

3. CAMPOS, D F., MATOS, A., and PINTO, A M., 'Multi-domain Inspection of Offshore Wind Farms Using an Autonomous Surface Vehicle', *SN Applied Sciences*, 2021.

4. KHALID, O., HAO, G., DESMOND, C., MACDONALD, H., McAULIFFE, F D., DOOLY, G., and HU, W., 'Applications of Robotics in Floating Offshore Wind Farm Operations and Maintenance: Literature Review and Trends', *Wind Energy*, 2022.

5. ŁEBKOWSKI, A., and KOZNOWSKI, W., 'Analysis of the Use of Electric and Hybrid Drives on SWATH Ships', *Energies*, 2020.

6. YETER, B., GARBATOV, Y., and SOARES, C G., 'Risk-based Maintenance Planning of Offshore Wind Turbine Farms', *Reliability Engineering & System Safety*, 2020.

7. STEHLY, T., BEITER, P., and DUFFY, P., '2019 Cost of Wind Energy Review', *National Renewable Energy Laboratory (NREL)*, 2020.

8. NETLAND, Ø., SPERSTAD, I B., HOFMANN, M., and SKAVHAUG, A., 'Cost-benefit Evaluation of Remote Inspection of Offshore Wind Farms by Simulating the Operation and Maintenance Phase', *Energy Procedia*, 2014.

9. KHALID, O., HAO, G., MACDONALD, H., COOPERMAN, A., DEVOY MCAULIFFE, F., and DESMOND, C., 'Cost-Benefit Assessment Framework for Robotics-Driven Inspection of Floating Offshore Wind Farms', *Wind Energy*, 2024.

10. HUANG, X., WANG, G., LU, Y., and JIA, Z., 'Study on a Boat-assisted Drone Inspection Scheme for the Modern Large-scale Offshore Wind Farm', *IEEE Systems Journal*, 2023.

11. SILVA, R J., LEITE, P N., and PINTO, A M., 'Multi-agent Optimization for Offshore Wind Farm Inspection Using an Improved Population-Based Metaheuristic', *IEEE International Conference on Autonomous Robot Systems and Competitions (ICARSC)*, 2020.

12. PENG, Z., SUN, S., TONG, L., FAN, Q., WANG, L., and LIU, D., 'Optimization of Offshore Wind Farm Inspection Paths Based on K-means-GA', *PLOS One*, 2024.

13. NORDIN, M H., SHARMA, S., KHAN, A., GIANNI, M., RAJENDRAN, S., and SUTTON, R., 'Collaborative Unmanned Vehicles for Inspection, Maintenance, and Repairs of Offshore Wind Turbines', *Drones*, 2022.

14. FOSTER, A J., GIANNI, M., ALY, A., SAMANI, H., and SHARMA, S., 'Multi-Robot Coverage Path Planning for the Inspection of Offshore Wind Farms: A Review', *Drones*, 2023.

15. KOU, L., LI, Y., ZHANG, F., GONG, X., HU, Y., YUAN, Q., and KE, W., 'Review on Monitoring, Operation and Maintenance of Smart Offshore Wind Farms', *Sensors*, 2022.

16. MARTINI, M., GUANCHE, R., LOSADA, I J., and VIDAL, C., 'Accessibility Assessment for Operation and Maintenance of Offshore Wind Farms in the North Sea', *Wind Energy*, 2017.

17. LUN, B K., and ALI, A., 'Analysis of Fin Stabilizer and Wave Effect on Free-Floating Semi-SWATH', *Journal of Transport System Engineering*, 2021.

18. ALIAKBARI, T., ADJAMI, M., and MOONESUN, M., 'An Experimental Study of Stabilizing Ordinary Fishing Nets (SOFNets) on a Stationary SWATH Ship Seakeeping Behavior Under Irregular Waves', *Ocean Engineering*, 2023.

19. GUAN, G., ZHUANG, Z., YANG, Q., WANG, P., and JIN, S., 'Hull Form Optimization Design of SWATH with Combination Evaluations of Resistance and Seakeeping Performance', *Ocean Engineering*, 2022.

20. GUTSCH, M., STEEN, S., and SPRENGER, F., 'Operability Robustness Index as Seakeeping Performance Criterion for Offshore Vessels', *Ocean Engineering*, 2020.

21. LI, B., QIAO, D., ZHAO, W., HU, Z., and LI, S., 'Operability Analysis of SWATH as a Service Vessel for Offshore Wind Turbine in the Southeastern Coast of China', *Ocean Engineering*, 2022.

22. DASKO, D., 'Feasibility of Hydrogen Powered <24m SWATH for ROV Deployment: Emissions Model Comparing the CO2 Emissions of SWATH and Monohull on the Route from Aberdeen to Future Windfarms Near Shetland Islands', *Numerical Seakeeping Report, CMDC2-SMMI, University of Southampton*, 2023.

23. TAUNTON, D., 'Feasibility of Seakeeping, Cost and GHG Efficiency of 24m SWATH Design for ROV Deployment', *Numerical Seakeeping Report, Project No: 10040003, University of Southampton*, 2023.

24. ANSYS, 'AQWA Theory Manual', *ANSYS*, 2022.

25. LEE, C M., 'Theoretical Prediction of Motion of Small-Waterplane-Area, Twin-Hull (SWATH) Ships in Waves', *David W. Taylor Naval Ship Research and Development Center*, 1976.

26. SUN, X S., YAO, C B., and YE, Q., 'Numerical Investigation on Seakeeping Performance of SWATH with Three-Dimensional Translating-Pulsating Source Green Function', *Engineering Analysis with Boundary Elements*, 2016.

27. BAITIS, A., COWELL, J., CROSSLAND, P., DANIEL, A., HOLCOMBE, F., KRIKKE, M., PATTISON, J., PIGEAU, R., RICH, K., and WERTHEIM, A., 'Generating and Using Human Performance Simulation Data to Guide Designers and Operators of Navy Ships: Two Large Multinational Programs', *Proceedings of RINA International Conference on Seakeeping and Weather*, 1995.

28. DET NORSKE VERITAS (DNV), 'Modelling and Analysis of Marine Operations', *Recommended Practice DNV-RP-H103*, 2011.

10. AUTHORS BIOGRAPHY

Suleyman Aykut Korkmaz is a Research Fellow at the University of Southampton in the Maritime Engineering Group, Department of Civil, Maritime, and Environmental Engineering. His expertise spans marine engineering, with a focus on hybrid propulsion systems, the integration of fuel cells and Li-ion batteries, the use of alternative fuels, and the decarbonization of maritime transportation.

Stephen Turnock, FRINA is Professor of Maritime Fluid Dynamics and currently Head of School of Engineering. His expertise is in the synthesis of computational, experimental, and analytical tools to maritime research challenges including maritime robotics and autonomy, performance sport, marine renewables, ship energy efficiency and decarbonisation.

Dominic Taunton is an Associate Professor at the University of Southampton. His research focuses on enhancing the design and operation of maritime vessels, particularly high-speed craft, such as semi-displacement catamarans. He specializes in experimental hydrodynamics, integrating human factors into ship design, and improving seakeeping and fuel efficiency of autonomous surface vehicles. Additionally, he has worked on performance sports engineering, adapting maritime research methods to sports like rowing and sailing.

Martyn Prince holds the position of Principal Research Engineer at the Wolfson Unit for Marine Technology and Industrial Aerodynamics at the University of Southampton. His primary areas of expertise involve aerodynamic and hydrodynamic experimental testing and vessel performance prediction.

Neil Timouth, CEO at ACUA Ocean. Neil was a founding member of the GEBCO-NF Alumni team that won the Shell Ocean Discovery XPRIZE competition for autonomous ocean mapping. Neil was also formerly the Chief Operating Officer (COO) of SEA-KIT International, a leading USV manufacturer, where he secured £3m of investment and funding, growing the company from a turnover of £50k in 2019 to over £3m in 2020.

Mike Tinmouth, Chief Operating Officer at ACUA Ocean. Mike has spent the last decade advising high-growth tech startups. BSc Law degree, major in Maritime Law and an MSc in Risk and Security Management. Mike mentors the PUBLIC GovStart accelerator, is a lecturer and business coach on the Royal Navy's innovation program, The Percy Hobart Fellowship.

EXPLORING THE FEASIBILITY OF AUTONOMOUS FERRIES FOR URBAN MOBILITY

S Eriksen, S V Hansen, and **A Orlovic**, VesOPS, DK

SUMMARY

Autonomous waterborne passenger transport is expected to revolutionize urban mobility, addressing congestion, pollution, and the need for sustainable solutions. This paper examines an automated and uncrewed passenger and bicycle ferry between Vindeby and Svendborg, Denmark, identifying three critical challenges: legal, technical, and commercial feasibility. Interviews with the Danish Maritime Authority reveal no fundamental legal barriers, though stringent safety and regulatory requirements must be met. Technologically, the ferry shows promise, but continuous operation without human oversight remains unproven. Commercially, the study highlights significant hurdles, with operational expenses nearly double the projected revenue, even with full automation.

1. INTRODUCTION

Urban mobility in many cities today is dominated by road transport, but as the focus increasingly shifts to sustainability, climate, and enhanced liveability, new transport concepts are being introduced. Historically, waterways served as the main form of transport, and many cities are built around rivers, fjords, or canals. Advances in electrical propulsion and storage, along with automated navigation—colloquially referred to as autonomous technology—may reintroduce waterways as a viable means of urban transport.

The introduction of small automated and uncrewed passenger ferries, or the vision of such a service, is emerging in multiple places, such as the MF Estelle in Stockholm [1], Greenhopper in Aalborg [2], and MilliAmpere 2 in Trondheim [3]. However, these services have so far been continuously manned and monitored by an onboard operator.

This paper is based on the findings of MARS sub-project 1.1e-a [4], which assesses the practical, legal, and commercial feasibility of realizing and operating a fully automated and uncrewed passenger ferry between Vindeby and Svendborg in Denmark.

2. GOAL AND SCOPE

The goal of the work described in this paper is to investigate the feasibility of introducing small automated and uncrewed passenger ferries in commercial operation. The purpose of the paper is to gauge the gap between the realization of this goal and the current technological capabilities, legal limitations, and commercial viability.

This study examines a case of a small automated and uncrewed ferry that can carry passengers and bicycles in a specific location. The choice of a case study as a method provides a solid foundation for exploring specific real-life issues and potential barriers to this type of operation. While the specific conditions of the study are unique to this case, they are still comparable to many others. It is believed that the applicability of this study and its methods extends to many other similar projects.

3. METHODOLOGY AND DATA

The case chosen for this study is detailed in Section 3.1. Many variables will be similar to those of existing manned vessels, but because of the novel nature of the concept, there are still considerable unknowns. To identify these unknowns, this study uses a commonly used method: the four quadrants, as visualized in Figure 4. This process is described in Section 3.2. The identified unknowns are explored in Sections 4, 5, and 6. Appropriate data to explore these unknowns are identified and analysed using suitable methods.

3.1 CASE

The case chosen for this analysis is a fully automated and uncrewed passenger and bicycle ferry operating between Vindeby and Svendborg in Denmark. This case is selected because such a service is specifically proposed in Svendborg Kommune's development strategy [5]. Additionally, the case fits well within the scope of the MARS project [4], under which this sub-project is undertaken.

In the following sections, the proposed route is detailed, the proposed ferry and needed shore terminal are described, and the operational scenario chosen for this study is outlined.

3.1 (a) Route

The proposed route, as shown in Figure 1, from Vindeby to Svendborg Port is approximately 550 meters in length, giving an estimated transit time of 7 minutes. The Svendborg-Vindeby area is a popular sailing destination and also serves a smaller volume of commercial shipping traffic. During the summer, the waters are especially busy with leisure crafts, kayaks, paddleboarders, and the like. The area is well sheltered from swell but does experience noticeable tidal and surface currents.

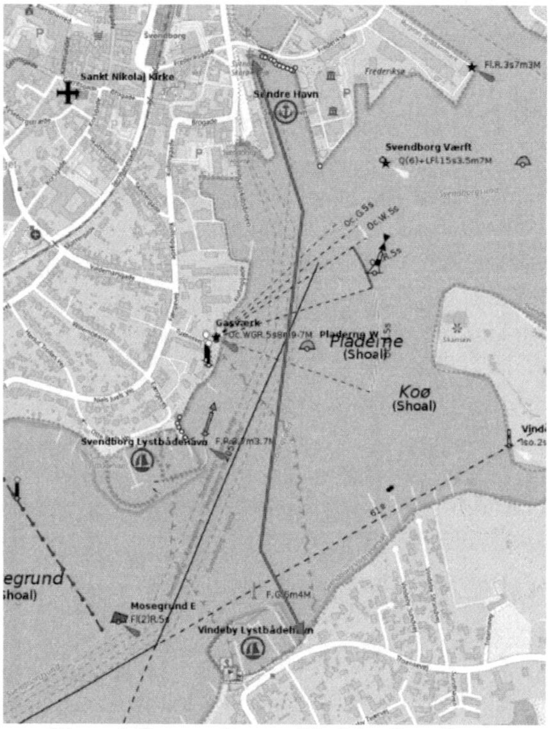

Figure 1: Proposed routes Vindeby-Svendborg.

3.1 (b) The Ferry

The proposed ferry is a double-ended catamaran similar in design to that shown in Figure 2. The ferry will have reversible propulsion and will be able to operate in both directions. The propulsion will be electrically powered by an onboard battery bank. The ferry will charge at a charging facility on the dedicated berths.

Figure 2: Concept for proposed ferry design [6].

The proposed ferry will accommodate up to 25 passengers as well as bicycles, baby strollers, etc. The ferry will be equipped with all hardware, sensors, and systems needed to enable fully automated operation and navigation, as well as remote monitoring and operation. However, the ferry will also have the facilities to accommodate an onboard operator.

3.1 (c) Land Facilities

The proposed ferry will have dedicated berthing facilities at both ends of the route. A pontoon berth similar to the one shown in Figure 3 is proposed. This design is considered a cost-effective solution and can be integrated into most existing infrastructure. The pontoon design will adapt to changes in tide and facilitate easy access to the ferry for pedestrians, bicyclists, wheelchair users etc.

Figure 3: Proposed land facility (https://www.zeabuz.com/images/).

3.1 (d) Operational Scenario

The ferry is intended to serve, among others, daily commuters and must therefore operate daily during normal working hours. To be valuable to tourists and the local community, the ferry should also operate in the evenings, on weekends, and during holidays. The proposed operations schedule is:

- Monday to Friday: 0700-2000

- Saturday and Sunday: 0900-1900

3.1 (e) Remote Monitoring and Operation

The ferry is designed to operate without an onboard operator after a two-year run-in period of permanently manned operation. In the uncrewed and automated mode, a Remote Operations Centre (ROC) will remotely monitor and control the ferry, with operators able to take over control if necessary. The assumption is made that an ROC will be available, and its design and operational costs are not detailed in this study.

Remote monitoring allows for the simultaneous oversight of multiple ships, depending on navigational needs. ROCs are a new concept, and there is limited experience regarding the number of ships an operator can effectively monitor. Given the complexity of the proposed ferry service, it is assumed that each operator can oversee a maximum of three ships at a time in this scenario.

While ROC operator salaries are expected to be similar to those of onboard operators, additional costs such as office space, equipment, and communication must be considered. Tasks that require physical presence on the ferry, such as inspections, cleaning, and equipment management, will still need onsite intervention, leading to further expenses.

No specific cost estimates for ROC operations or onboard interventions are provided here. Given the assumptions, it is projected that salary expenses will be reduced by half when the ferry shifts to remote monitoring and control compared to onboard operation.

3.2 IDENTIFYING UNKNOWNS

All development projects contain uncertainty but for a novel project such as the one described in this study, the unknowns are especially abundant. Numerous potential unknowns could bring the described project to a stop, cause it to result in financial loss, or even worse, bodily harm to passengers, crews, or others. Eliminating uncertainties and potential failures is nearly impossible as evidenced by e.g. the Viking Sky incident [7] or the Challenger space shuttle disaster [8]. However, many unknowns can and should be identified.

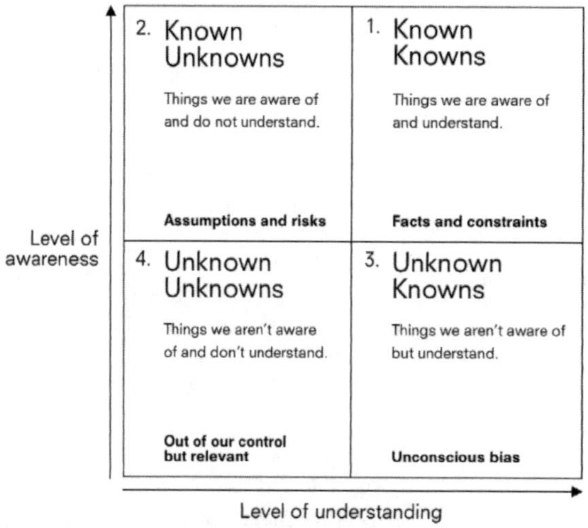

Figure 4: Identifying knowledge gaps (https://shorturl.at/bJ6m0).

As depicted in Figure 4, the method aims to increase awareness or understanding of unknown issues, ideally achieving both. Some unknowns can be identified, examined, and understood, while others may only allow for gauging potential outcomes or monitoring. In such cases, this report aims to delineate knowledge gaps and suggest further investigation points. Additionally, there are known issues that may be consciously or unconsciously neglected. Everyone is subject to biases, but through examination, these biases can hopefully be minimized. The most dangerous unknowns are those that are completely unforeseen. While such uncertainties can never be fully eliminated, they can be mitigated through experience and by learning from others.

Sections 3.2 (a) to (d) describe the identification of knowns and unknowns using this method, focusing on issues believed to have the greatest potential impact.

3.2 (a) Known Knowns

- Ferry purchase cost: Estimated with confidence.
- Autonomy package cost: Quote received.
- Land facility cost: Estimated with confidence.
- Operation costs: Reference costs available.

3.2 (b) Known Unknowns

- Transport demand uncertainties include:
    - Historical changes since the last ferry service in 1966.
    - Lack of direct transport data between proposed destinations.
    - Uncertain commuting patterns (car, bus, bicycle).
    - Uncertain willingness to travel to and from the ferry stops.
    - Potential impact on transport demand arising from the new service.
    - Novelty effect on passenger numbers.

- Legal feasibility uncertainties:
    - Current legal restrictions on uncrewed vessels.
    - Compliance challenges with passenger safety rules.
    - Requirements and procedures for demonstrating vessel capabilities.

- Technical feasibility uncertainties:
    - Limited commercial success of uncrewed and automated ships.
    - Operational reliance on onboard operators in existing projects.
    - Challenges in accurately assessing technological readiness.

3.2 (c) Unknown Knowns

- Biases such as optimism and confirmation biases.
- Economic interests influencing project assessments.

3.2 (d) Unknown Unknowns

- Unidentified potential barriers not covered in this analysis e.g. opposition from locals or labour unions.

3.3 IDENTIFIED UNKNOWNS

From the analysis in Section 3.2, three categories require further exploration: legal-, technical-, and commercial feasibility. Detailed data collection and analysis methods for each category are described in Sections 4, 5, and 6.

4. LEGAL FEASIBILITY

Maritime affairs are in Denmark administered and regulated by the Danish Maritime Authority (DMA) [9]. The DMA is therefore the logical source for answers to the unknowns regarding safety aspects and legal feasibility identified in section 4.1. Through the existing contact between the MARS project and the DMA a line of communication was established. Representatives from the DMA generously agreed to consider and answer the questions generated during this study regarding the safety aspects and legal feasibility of implementing and operating a small automated and uncrewed ferry such as described in section 3.1.

4.1 THE INTERVIEW

A list of topics for discussion was outlined and sent to the DMA before the interview. The questions dealt with the following subjects:

- Existing legislation on uncrewed and automated operation
- What changes (if any) are needed to allow for such an operation?
- Which technical systems must be in place to allow for such operation?

- How can applicants document compliance?
- Which requirements are put to ROCs in terms of:
    - Physical location
    - Competence and training of operators
- Can special restrictions be imposed on the ferry to easier enable uncrewed and automated operation?

The interview was held at DMA's offices in Korsør on July 04, 2024. The interviewers were Charlotte Kirkegaard Flugt and Stig Eriksen from MARS and the interviewees from DMA were Jakob Bang, Lasse Grøn Christensen, Torsten Arndt Olsen, and Ronny Marstrand Hansen. The interview was conducted in Danish and was recorded for internal reference.

4.2 RESULTS

The interview explored whether the Danish Maritime Authority (DMA) could approve the uncrewed and automated operation of a passenger ferry under existing regulations. In short, the DMA do not as such see any regulatory barriers in the given case (short distance, up to 25 pax within territorial waters) for which regulatory solutions towards and approval cannot be found. The crucial aspect will be to ensure and demonstrate that the safety of passengers, vessel and navigation will be satisfactorily and at least equivalent to the safety level of a traditional concept. In case of specific rules barring such operations within Denmark's national territorial waters, the DMA have the authority to make exemptions in specific cases if safety and environmental conditions are met. However, approval hinges on a detailed description and demonstrating the ferry's ability to perform tasks satisfactorily. This include the ability to maintain navigational safety in normal operation, incl. the ability to deal with the complex traffic in Svendborgsund and strong currents in places, as well as emergency situations, where passenger safety must be at the same level as in manned operations The interviewees emphasized this assessment applies solely to this case and only within national waters. In general, the DMA handle enquiries within MASS (Maritime Autonomous Surface Ships) on a case-by-case basis and in cooperation with the industry, the DMA will contribute through a constructive approach to find ways for approvals of safe, secure and environmentally friendly solutions. Further development of the concept and a possible future approval would therefore need to be based on a number of fundamental safety thresholds, and that the operator can demonstrate an equivalent level of safety.

Regarding future regulatory changes, no immediate changes are seen as necessary for the approval of this specific case, but equivalences and exceptions would have to be made. However, the DMA generally expects that changes will be made in the future to ease the approval process, when autonomous operations become more widespread. Changes to national regulations will be based on experiences gained through a number of concrete projects.

In terms of technical requirements, the DMA applies a risk-based approval approach and an equivalence principle: the safety standards for passengers on an uncrewed ferry must match or exceed those on a crewed ferry. Compliance with International Regulations for Preventing Collisions at Sea (COLREG) is crucial, necessitating practical demonstrations rather than mere rule application due to their interpretative nature.

The DMA also discussed Remote Operations Centers (ROC), which would also be assessed as part of the DMA approval. Currently, there are no specific rules governing ROC design and capabilities, but robust communication systems and operational contingency plans are key considerations.

Training requirements for ROC operators are yet to be defined. The DMA explains that there will likely be many overlaps with the requirements for the existing training of conventional seagoing operators, but there will likely also be differences. Additional system-specific training may be needed whereas other current training requirements may not be necessary.

The DMA did not foresee any special designations or restrictions specific to the proposed ferry for enabling uncrewed and automated operation. They clarified that designating a ferry as a virtual cable ferry, for example, would not grant it special rights under existing regulations.

4.3 REMARKS ON THE RESULT

Currently, there are no commercially operational fully uncrewed and automated passenger ships, and the Danish Maritime Authority has not yet reviewed an application for such a vessel. The interview provided by the DMA is based on a theoretical scenario, lacking sufficient detail for an actual approval process. Despite these limitations and numerous uncertainties, the DMA conducted a comprehensive analysis of the questions posed.

Several issues lack specific regulatory guidance or precedent due to the novelty of the concept, and a risk based approval process (following the principles set out in the MSC.1/Circ.1455) would be applied. Nonetheless, there are no impediments within the current legislation that would preclude regulatory solutions from being identified, and the DMA is confident that approval could be granted if the conditions outlined above are met.

Key interview points include:

- Regulatory Flexibility: The DMA is able to find regulatory solutions for approvals of an autonomous operation as described in this case given that safety and environmental conditions are met.
- Safety Requirements: The uncrewed ferry must ensure passenger safety equivalent to crewed ferries, especially in emergencies and rescue operations.
- COLREG Compliance: Automated ferries must adhere strictly to COLREGs, and rules will not change to accommodate automated and uncrewed operation.
- Testing and Oversight: The DMA requires a trial period with continuous monitoring as well as full-scale tests of accident scenarios to verify compliance and safety.
- Remote Operations Centres: An assessment of the ROCs would need by part of the DMA approval, considering factors like reliable communication, safe design and ergonomics, and local intervention capacity etc. In general, it is expected that ROCs shall be placed in the territory of the Flag state and in this case Denmark.
- Operational Standards: No special shortcuts or designations (e.g., virtual cable ferry) will simplify compliance; all technical and safety criteria must be fully met.

In conclusion, while there are no fundamental legal barriers to small, short distance uncrewed and automated passenger ferries within Danish territorial waters, fundamental safety and regulatory compliance requirements must be met to obtain DMA approval.

5. TECHNICAL FEASIBILITY

Automated and uncrewed ship operations have primarily been limited to functional testing stages. While initial findings are generally positive, none of these projects have yet transitioned to full commercial operation. Recent developments include services like CallBoats in Helsinki [10] and ZeaBuz in Stockholm [11], which operate with onboard operators continuously monitoring operations. However, projects such as the Greenhopper [2] in Aalborg, have encountered difficulties progressing from development to operational stages, though publicly available information doesn't specify the technical feasibility's exact role in these outcomes.

Currently, operational performance data are scarce for these systems. Metrics such as object detection success, collision avoidance accuracy, system uptime, near-miss incidents, and human operator interventions would have been interesting to examine but have not been available for this study. This lack of data likely stems from the systems' recent deployment history and commercial considerations, as private companies developing these technologies may withhold data to maintain competitive advantage. Furthermore, projects that were not fully realized may also be hesitant to share negative results (section 3.2.3).

5.1 EXPERT ASSESSMENT

In assessing technical feasibility, this study relies on the expertise of Lasse Skriver Møller (LSM), Co-CEO and Partner at DanaDynamics (DD), a technology firm specializing in autonomous solutions for unmanned surface vessels since 2018 [12]. DD, a partner in the MARS project, has generously provided its expertise for this study.

LSM possesses in-depth knowledge of DD's products and a comprehensive understanding of their competitors' offerings. As a seasoned expert in the field, he also navigates the broader technical challenges, limitations, and potential barriers associated with automated and uncrewed ship operations.

The expert assessment took the form of a semi-structured interview held between Stig Eriksen and Lasse Skriver Møller in Svendborg on June 19, 2024.

5.2 RESULTS

The interview commenced with an exploration of DanaDynamic's core interests and current business focus. Lasse Skriver Møller outlined DD's primary commitment to developing autonomy systems for marine surface survey drones, with some attention also directed towards autonomous systems for naval applications. While DD does not currently offer a system precisely aligned with the requirements for the ferry in question, LSM indicated it as an area of future interest and development for the company.

LSM outlines the following areas of technological development to be addressed: automated navigation, collision avoidance, auto-docking, passenger handling and safety, and highlighted connectivity as an additional critical concern.

Automated navigation is feasible according to LSM, and this feature is already largely incorporated into DD's software. He noted that similar systems are already market-ready for automated navigational planning and execution, capable of adjusting to environmental changes during sailing.

Regarding automated collision avoidance, LSM acknowledged that current systems, including DD's, are not fully compliant with the COLREGs. While capable in many scenarios, the challenge lies in interpreting ambiguous COLREG terms and predicting vessel intentions accurately. LSM sees potential in advancements through artificial intelligence (AI) but stresses the need for more training data specific to marine applications.

Auto-docking, comparatively, was described as a more straightforward task. LSM emphasized that auto-docking may not be possible for all vessels, but commercially operational products are available on the market.

On passenger handling and safety, LSM raised concerns about the open-deck configurations common in smaller ferries, highlighting potential issues in emergency situations where onboard personnel are absent.

Connectivity emerged as a critical technical hurdle, essential for maintaining remote operational oversight. LSM underscored the complexities of ensuring robust, secure ship-to-shore communication amidst potential vulnerabilities like hacking and signal interference.

In comparing challenges with autonomous ships to self-driving cars, LSM noted unique difficulties in maritime settings, particularly in unstructured near-shore environments.

Regarding the commercial viability of automated ferry operations as described in this study, LSM expressed scepticism due to uncertainties in the capabilities and cost-effectiveness of ROC centres.

In LSM's opinion, the described scenario is not technically possible now. He is hesitant to give a definite answer but suggests a realistic timeline of ten years or more.

5.3 REMARKS ON THE RESULT

Lasse Skriver Møller's assessment, based on his extensive expertise in the field, provides valuable insights. While other experts may hold differing opinions and experiences, LSM acknowledges he is not the definitive authority on the subject, nor is there one universally recognized. Bias is inherent, as discussed in section 3.2.3, yet LSM's insights align closely with current research and findings available to the authors at the time of writing.

Key points from the expert interview are:

- Automated Navigation Planning: Automated navigation planning, execution of the sail plan, and auto docking are technically possible and can be implemented in commercial operations today.
- Automatic Collision Avoidance: Automatic collision avoidance is still not technically possible, but LSM believes it will be. Significant work is being done on this subject.
- Passenger Handling and Safety: Passenger handling and safety is problematic. This subject has not received much attention, and the challenges and solutions remain unclear.
- Financial Benefit: The financial benefit from automation/autonomy systems on ships may not come from reduced operational costs, but from increased safety.

In summary, while some aspects of the proposed operation are feasible, fully automated and uncrewed operation, as suggested here, is assessed not to be technically possible today. LSM estimates that achieving this scenario could realistically take ten years or more. LSM acknowledges that this estimate is subjective, and others argue that such a project has been technically feasible for years [13]. Regardless of these perspectives, automated ships have not yet been demonstrated to operate without a human operator present for extended periods.

6. COMMERCIAL FEASIBILITY

Automated and uncrewed operation of ships is not a well-established concept and reference data that completely mirrors that of the case described in section 3.1 is therefore not readily available. Apart from the specifics relating to the automated and uncrewed operation, however, the proposed vessel is largely conventional, and for this, good reference data is accessible.

Two sources that have themselves explored projects similar to that described in this study were kind enough to share their data. A condition for sharing the data was that it would not be distributed further, and therefore it cannot be reproduced directly here. The shared data have instead been used by the authors as a reference for the cost estimates presented in this section.

The estimation of the passenger potential which, as described in section 3.2, is one of the large unknowns in this project is presented in section 6.1. Section 6.2 presents the calculations of the construction costs related to this project. The calculation of the operational expenses and the values and assumptions that form the basis for this calculation is shown in Section 6.3.

6.1 REVENUE

The proposed revenue source is ferry ticket sales, hinging on passenger potential, as discussed in section 3.2.2. This study employs Jensen et al. [14] population-based method to estimate passenger demand. Jensen et al. [14] analysed a small, crewed ferry in Aalborg and adapted their approach from a Copenhagen urban water transport study. In Copenhagen, they estimated a total passenger potential of 720,763, evenly distributed among commuters, local leisure travellers, and tourists (240,253 each). Applying a comparative density approach, Jensen et al. [14] scaled down these figures for Aalborg.

Both studies utilize population and workplace densities within a 500-meter radius around ferry stops. This study extends to a 1000-meter radius due to the ferry's bike-friendly nature, assuming bicyclists travel further than pedestrians.

Population and workplace density data for 500-meter and 1000-meter radii around Vindeby Marina and Svendborg Port were provided by Svendborg Kommune. To avoid overlap, adjustments were made to exclude overlapping areas between Vindeby and Svendborg/Taasinge, facilitated by Svendborg Kommune. Data presented in Table 1 were received on May 28, 2024, from Svendborg Kommune's database.

Table 1. Population and workplace density.

	Population		Workplaces	
	500m	1000m	500m	1000m
Vindeby Marina	844	1776	21	502
Svendborg Port	1472	5335	2652	9106

The number of potential passengers decreases with distance to the station/ferry stop [15]. Passenger potential for the area outside the 500-meter zone is therefore likely much less than that within the zone. To compensate for that, the population and workplace density for the zone between 500 and 1000 meters is multiplied by a factor of 0,2 in this study. As the 1000-meter zone includes the 500-meter zone, the values from the second must naturally be subtracted from the first. Table 2 shows the calculation of the corrected totals.

Table 2. Corrected total.

	Population	Workplaces
Vindeby Marina		
Radius 500 m	844	21
Radius 1000 m - corrected*	186	96
Svendborg Port		
Radius 500 m	1.472	2.652
Radius 1000 m - corrected*	773	1.291
Corrected total	**3.275**	**4.060**

**(radius 1000m - radius 500m)·0,2*

As in Jensen et al. [14], the corrected total population and workplace density values are compared to reference values from the Copenhagen case study. It is shown by the resulting factor that the population, for example, is 8.2 times higher in the Copenhagen study than in the Svendborg-Vindeby case. The passenger potential from the Copenhagen case, 240,253 for each category, is then divided by the calculated factor to estimate the passenger potential for the present study. The resulting values are shown in Table 3.

Table 3. Passenger potential.

	Population	Workplaces
Corrected total	3275	4060
Copenhagen reference values	26885	68226
Factor	8,2	16,8
Passenger potential	**29.266**	**14.297**

Jensen et al. [14] also include the passenger category 'Tourists'. Unfortunately, data for the number of hotel stays in the area on which this value for this category is based have not been available for this study. Instead, it is assumed that the proportion of tourists to commuters plus local leisure travellers is the same as in Jensen et al. [14]. In the Aalborg case, the used passenger potential values are Commuters: 52.644, Local leisure travellers: 19.632, and Tourists: 26.695. For the present case, this corresponds to a tourist passenger potential of 16.090. Combined across all three categories the total passenger potential used in this case is therefore 59.654 per year. At an assumed ticket price of 20 DKK (€2,67) excl. moms the annual revenue will be 1.193.171 DKK (€159.076).

The passenger potential calculated here is believed to be the best possible estimate based on the available data and resources. However, the estimate is based on many assumptions and approximations. Passenger potential still presents considerable uncertainty for the case in this study and a more thorough study be done if it is decided to proceed with the project. The estimate presented here likely presents a "best-case" scenario.

6.2 CONSTRUCTION COSTS

The purchase, build, and establishment costs for the ferry and associated land facilities can, as mentioned in section 3.2, be estimated with a good degree of certainty. Table 4 shows the projected construction cost budget.

Table 4. Construction cost budget.

	DKK
Ferry and related equipment	
Ferry	7.458.800 kr.
Autonomy package	1.827.406 kr.
Subtotal	**9.286.206 kr.**
Shore facilities	
Docking pontoons	559.410 kr.
Electrical charging system	1.342.584 kr.
Project management	200.000 kr.
Ticketing system	300.000 kr.
Subtotal	**2.401.994 kr.**
Total	**11.688.200 kr.**

The cost of the autonomy package is based on an actual offer, whereas the remaining values are based on informed estimates which align well with the reference values. A total build cost of just below 12 million DKK (approx. €1,5 million) is projected. This aligns well with the reference values and is believed to be a realistic budget.

6.3 OPERATIONAL EXPENSES

Projected operational expenses for the first three years are detailed in Table 4. Initial ticket sales revenue is expected to ramp up gradually over this period.

The salary budget is determined by the operational schedule outlined in section 3.1.4. Including startup and shutdown, worktime totals 92 hours per week or 4.797 hours annually. Given a full-time equivalent of 1.571 hours per year [16], continuous manned operation requires three full-time positions. At an hourly rate of 250 DKK (€33.3) plus benefits, the annual salary cost in year one totals 1.5 million DKK (€200.000), increasing by five per cent annually. By year three, transitioning to automated operation is expected to halve these expenses (section 3.1.5).

Insurance costs are initially estimated at 75.000 DKK (€10.000) for manned operation but are projected to double when shifting to automated operation due to its novel nature. The remaining expenses are anticipated to remain constant. The annual autonomy package subscription is based on a specific quote, while other figures align with reliable estimates.

Table 5. Operational expenditure budget.

Revenue	Year one	Year two	Year three
Passenger phase-in	66%	90%	100%
Ticket sale	787.427 kr.	1.073.764 kr.	1.193.171 kr.
Variable costs			
Salaries	1.500.000 kr.	1.575.000 kr.	826.875 kr.
Electricity	300.000 kr.	300.000 kr.	300.000 kr.
Subtotal	**1.800.000 kr.**	**1.875.000 kr.**	**1.126.875 kr.**
Fixed costs			
Maintenance ferry	200.000 kr.	200.000 kr.	200.000 kr.
Maintenance shore facilities	75.000 kr.	75.000 kr.	75.000 kr.
Autonomy package yearly subscription	544.514 kr.	544.514 kr.	544.514 kr.

Survey and lifesaving equipment	100.000 kr.	100.000 kr.	100.000 kr.
Insurance	75.000 kr.	75.000 kr.	150.000 kr.
Subtotal	**994.514 kr.**	**994.514 kr.**	**1.069.514 kr.**
Total	**- 2.006.919 kr.**	**- 1.795.582 kr.**	**- 1.003.150 kr.**

The budget shows a total deficit of about 1,8 to 2,0 million DKK (approx. €250.000) for the first two years of manned operation and a deficit of about 1,0 million DKK (approx. €135.000) for year three when the ferry is automated and uncrewed. Over the first three years, a total deficit of 4,8 million DKK (approx. €640.000) is projected for the operation.

If a project as described in this study were to be realised it is possible that a large part of the construction/establishment cost could be secured through external funding. The projected cost for the operators would therefore be those presented in Table 5. However, were the operators also to fund the construction/establishment themselves, financial costs must be included as shown in Table 6. A twenty-year lifespan of the ferry and related equipment is assumed, and a conservative effective interest rate of six per cent is used.

Table 6. Operational expenditure budget incl. financial costs.

Depreciation over twenty years and financial costs			
Ferry and equipment	464.391 kr.	464.391 kr.	464.391 kr.
Land facilities	120.116 kr.	120.116 kr.	120.116 kr.
Financial costs	701.409 kr.	666.339 kr.	631.268 kr.
Subtotal	**1.285.917 kr.**	**1.250.846 kr.**	**1.215.776 kr.**
Total from Table 5	**- 2.006.919 kr.**	**- 1.795.582 kr.**	**- 1.003.514 kr.**
Result before tax	**- 3.292.363 kr.**	**- 3.045.969 kr.**	**- 2.218.480 kr.**

A yearly cost of 1,2 to 1,3 million DKK (approx. €165.000) for financing and depreciation is projected. Added to the results for the operational budget this results in a total deficit over three years of just over 8,5 million DKK (€1,1 million).

6.4 REMARKS ON THE RESULTS

The results of the analysis of the commercial feasibility presented in sections 6.1, 6.2, and 6.3 are visualised in Figure 5. Overall, it is clear that the revenue fails to meet the projected costs of operating and financing the ferry. The passenger potential is too low to generate enough income to match the expenses.

The first two years show the largest deficits due to high salary costs and a gradual phase-in of passenger numbers. However, the deficit remains significant even after transitioning to automated and uncrewed operations. Almost twice as many passengers as projected must use the ferry for the revenue to meet the operational expenses when in automated and uncrewed operation in year three. If financial expenses are also to be covered by ticket sales, almost three times as many passengers as projected must use the ferry.

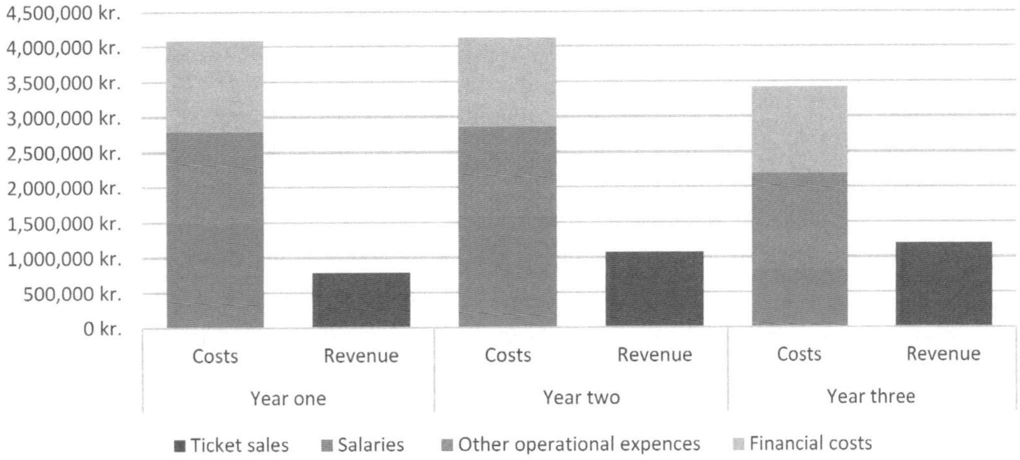

Figure 5: Costs versus revenue.

The described scenarios and preconditions are based on the best available data but also on many informed assumptions. Changes to these may alter the specifics of the calculations but overall, the results show that there is a wide gap between revenue and cost. As a commercial venture, the analysis shows that the proposed case is not feasible if revenues are to come from ticket sales alone.

7. CONCLUSIONS

This study assesses the potential for an automated and uncrewed ferry service connecting Svendborg and Vindeby in Denmark, carrying up to 25 pax, focusing on legal, technical, and commercial feasibility.

Legally, there are no regulatory barriers in the given case for which regulatory solutions towards an approval cannot be found. The crucial aspect will be to ensure and demonstrate that the safety of passengers, vessel and navigation will be satisfactorily and at least equivalent to the safety level of a traditional concept allowing the Danish Maritime Authority (DMA) flexibility to grant exemptions if safety standards equivalent to traditional crewed ferries are met. Compliance with International Regulations for Preventing Collisions at Sea (COLREG) is mandatory without exceptions for automated vessels. The DMA requires operational testing with continuous monitoring as well as full-scale tests of accident scenarios and mandates an assessment of the Remote Operations Centres (ROC) as part of the DMA approval.

Fully automated and uncrewed ships face significant technical challenges. While commercial applications can handle navigation planning, sailing plan execution, and auto-docking, automatic collision avoidance and passenger safety are still problematic. Despite varying assessments, this case study concludes that commercial implementation of fully automated and uncrewed operation of passenger ferries is realistically over a decade away.

Commercially, the analysis reveals substantial deficits due to projected revenue falling short of operational costs. Even with automation in the third year, deficits persist, necessitating nearly triple the projected passenger numbers to break even. Despite detailed analysis, the study concludes that the proposed automated and uncrewed ferry operation is not financially viable solely on ticket sales.

In summary, while legally feasible with strict compliance, technically achievable in part, and promising conceptually, significant challenges must be overcome before commercial operation of automated and uncrewed ferries is realistic.

8. ACKNOWLEDGEMENTS

The authors would like to acknowledge The Danish Maritime Authority, DanaDynamics, Svendborg Kommune, Fynbus, ZeaBuz, and Port of Aalborg for their invaluable contribution to the study on which this paper is based.

9. REFERENCES

1. RANTORP, C G., 'Nordmænd bag førerløs by-færge i Stockholm har sat kursen mod København | MobilityTech (PRO)'. Accessed: May 23, 2024. [Online]. Available: https://pro.ing.dk/mobilitytech/artikel/nordmaend-bag-foererloes-faerge-i-stockholm-har-sat-kursen-mod-koebenhavn

2. DJURSING, T., 'Den lille grønhoppe kommer aldrig ud at sejle i Aalborg Havn | Ingeniøren'. Accessed: May 23, 2024. [Online]. Available: https://ing.dk/artikel/den-lille-groenhoppe-kommer-aldrig-ud-sejle-i-aalborg-havn

3. MARITIME, B., 'VESSEL REVIEW | MilliAmpere 2 – Norwegian university develops autonomous electric ferry demonstrator', Baird Maritime. Accessed: May 23, 2024. [Online]. Available: https://www.bairdmaritime.com/passenger/ferry/vessel-review-milliampere-2-norwegian-university-develops-autonomous-electric-ferry-demonstrator/

4. 'MARS Denmark', Simac.dk. Accessed: Jun. 25, 2024. [Online]. Available: https://marsdenmark.dk/

5. Svendborg Kommune, 'Udviklingsstrategi 2023-2035'. Aug. 29, 2023. [Online]. Available: https://udviklingsstrategi.svendborg.dk/media/5525/udviklingsstrategi-2023-2035-fra-mere-til-bedre-fremtidens-svendborg.pdf

6. MARQUART, A K., 'Stockholms autonome havnebus skal afgå hvert kvarter', Søfart. Accessed: Jun. 24, 2024. [Online]. Available: https://www.soefart.dk/article/view/898548/stockholms_autonome_havnebus_skal_afga_hvert_kvarter

7. Sjøfartsdirektoratet, 'Response to the NSIA report on the Viking Sky incident'. Accessed: May 27, 2024. [Online]. Available: https://www.sdir.no/en/news/news-from-the-nma/response-to-the-nsia-report-on-the-viking-sky-incident/

8. FEYNMAN, R., 'Report of the presidential commission on the space shuttle challenger accident', *Append. F*, p. 7, 1986.

9. 'Danish Maritime Authority'. Accessed: May 24, 2024. [Online]. Available: https://www.dma.dk/

10. KULTANEN, P., 'World's first autonomous boat comes from Helsinki', Forum Virium Helsinki. Accessed: Jun. 06, 2024. [Online]. Available: https://forumvirium.fi/en/release/worlds-first-autonomous-on-demand-boat-in-helsinki-could-turn-water-transport-carbon-neutral/

11. Zeabuz, 'Torghatten and Zeabuz Make History in Stockholm', Zeabuz. Accessed: May 27, 2024. [Online]. Available: https://www.zeabuz.com/torghatten-and-zeabuz-make-history-in-stockholm/

12. DanaDynamics, 'About us | DanaDynamics - Autonomous Systems'. Accessed: Jun. 06, 2024. [Online]. Available: https://danadynamics.com/about-us/

13. BREKKE, E F., 'Autonomous Shuttle Ferry Ferry in Trondheim'. *Trondheim*, Aug 15, 2017.

14. JENSEN, R., KRISTENSEN, J N., and KRISTENSEN, J., 'Projekt "Havnebus"', Aalborg Universitet, Center for Logistik og Samarbejde, Aug. 2017. Accessed: Jun. 03, 2024. [Online]. Available: https://vbn.aau.dk/files/265001599/Projekt_Havnebus_i_Limfjorden_faerdig_rapport.pdf

15. WARBURG, V., and RUE, I., 'Overlappende stationsoplande: Bestemmelse af passagerpotentialer', presented at the Proceedings from the Annual Transport Conference at Aalborg University, 2006.

16. OECD, 'Employment - Hours worked - OECD Data', theOECD. Accessed: May 27, 2024. [Online]. Available: http://data.oecd.org/emp/hours-worked.htm

9. AUTHORS BIOGRAPHY

Stig Eriksen holds the position of Energy Specialist at VesOPS. The Author has done research in the maintenance and operation of autonomous ships and holds a Ph.D. in Engineering from the University of Southern Denmark.

Søren Vinther Hansen holds the position of Director at VesOPS. The author has extensive knowledge of energy optimization in the maritime sector and holds a Ph.D. in Engineering from the Technical University of Denmark (DTU).

Andjela Orlovic holds the position of Vessel Performance Analyst at VesOPS. The author is responsible for data validation and holds a master's degree in Naval Architecture and Marine Engineering from the University of Belgrade.

Autonomous Ships 2024
20th November – 21st November 2024
Copenhagen, Denmark

IMPLEMENTING A NEW PLATFORM INTO THE U.S. NAVY: A SYSTEMS VIEW

S P Gallup, Naval Postgraduate School, USA

SUMMARY

The U.S. Navy needs to consider its capacity to confront threats. Alone, there are inadequate fleet assets. And even in alliances and cooperation with other nations this may not be enough. A concept of a "bi-modal" navy has been advanced [1] Kline, Jeffrey K, 2024] that adds smaller warships [2] and autonomous capabilities, working together to cover large portions of the Indo-Pacific. The Lightly Manned Automated Combat Corvette (LMACC) is a highly developed design concept that could answer the need for the advancing the bi-modal fleet. Although all the components are common in the U.S. navy, and industry, it is still a concept that can be sidetracked over implementation of new ideas. We take the position in this presentation that implementation is part of a wider system of systems, and that adherence to very long platform service lives affect nearly all part of this system.

NOMENCLATURE

LMACC	Lightly Manned Automated Combat Capability
PLA-N	People's Liberation Army-Navy
SLOCS	Sea Lines of Communication
PMW	Program Manager, Warfare
DON	Department of the Navy

1. INTRODUCTION

These are challenging times for defenders of the seas against belligerent nation states and non-nation state actors. Much is at stake as hostilities increase globally. The PLA-N is attempting a fait accompli across SLOCS in the Pacific, Iran works with proxies to contest the Red Sea, Russia is pushing its naval influence farther into blue waters mostly left at the end of the Cold War. War at sea concepts have also expanded, threatening undersea cables, and adding to other areas of risk around the globe. The U.S. Navy needs to consider its capacity to confront these threats. Alone, there are simply inadequate fleet assets as currently stands. And even in alliances and cooperation with other nations this may not be enough. A concept of a "bi-modal" navy has been advanced [1] that adds smaller warships [2] and autonomous capabilities, working together to cover large portions of the Indo-Pacific, with allied forces in the first and second Indo Pacific Island chains. For example, the Philippine fisheries and threats of invasion of Taiwan loom large to those in the region.

Platform service life is a key component limiting the inclusion of new capabilities that can be implemented quickly. U.S. Navy warfighting doctrine has an equally long "service life." This limits flexibility of surface fleet assets to emerging threats. The effect is that by the time a new ship is introduced to the fleet, its ability to be implemented successfully is diminished. Global dynamics unfold quickly.

We argue here for shorter service life warships, intended to not be updated in 30-year service lives, but replaced quickly, and meeting near term goals of implementation into the warfighting fleet. A systems view is taken such that the influences that act as feed-forward and feed-back create an unstable system of new ship design in time to be successfully included in the fleet and destabilizes U.S. and Allied combined arms capabilities.

2. CONCEPT AND NEW PLATFORM IMPLEMENTATION

2.1 SYSTEM FACTORS

A sense of urgency could affect the acceleration of change and diversification of portfolio. This is especially true when an existential threat exists. in development of new concepts and technology adapted to what is available for conducting war at sea. The adaptation of fishing vessels built in civilian yards to become the Flower class DEs continues to be an excellent example. An existential threat did exist (protection f convoys providing war materials to England during WW II). Two Hundred and eighty vessels were built during the war, adapting sonar and radar and new tactics to these small and cramped vessels. Their contribution had much to do with success of the convoy link from America, Canada and Great Britain. Without a clear existential threat what can be done to lessen the time needed to manifest what is truly needed when that threat *does* exist and employ technology evolution? Do we only respond to existential threats with completely new ideas? Even as technology advances are made, acquisition timelines are such that the time devoted to procuring new ships and revolutionary capabilities can be outpaced in the end by the development of new technology. What creates a sense of urgency? In his book, A Sense of Urgency [3], John Kotter explains that a true sense of urgency is rare; mainly because "it is not the natural situation. It must be created and recreated." So, the task of leading a team of people in a transformation at any level will often require an ability to create an atmosphere of urgency that can be embraced, and in turn bring about an atmosphere of achievement.

The U.S. DoD produced an Integrated Roadmap [4] where pressures to move to a different warfighting concept is the result of factors such as reductions in federal budgets, increasing complexity of the operational environment, rebalancing of U.S. forces towards the Pacific, unmanned technologies will continue to improve and cyber warfare will become a new warfighting domain. Although a roadmap, it has not created a specific concept or set of concepts that can be implemented across the operational domains of war at sea. Do we really have an effective autonomous and deterrent force in the Indo-Pacific? I would argue no. And the attempts at producing this capability (e.g., the U.S. *Sea Hunter* class is far from what is needed. The primary efforts around combat autonomy seem to be at the level of improved participatory control, and not in true combat autonomy. Thus, human control is still going to be needed, whether forward near the vessels, or back in the rear. This introduces the problem of communications resilience, which cannot be counted on in wartime, and opens the door to cyber warfare at the platform level. For the time being, we need humans at sea, on smaller ships with new concepts of operations. Hence the logic behind the LMACC. The primary problem here is that U.S. Navy investments do not go towards an obvious result, but continues to support the esoteric, long lead time, long platform life of the Arleigh Burke class DDGs and the new Constellation class FFGs. These are expensive, far too difficult to build in civilian yards at scale, and in the end will only provide a ship class of abut 20 vessels in the next 15 years. The Constellation has grown by 20 percent in weight, increased crew requirements and nearly doubled in price. Adapting a foreign ship calss to the U.S. Navy is proving to be a very difficult problem. In other words, this is not the answer to building "many" but once again represents the "few."

So far our "system" includes 1) agency (urgency), 2) king hull life, 3) crew requirements meeting older design criteria, 4) lack of an evolutionary path to include shorter vessel service life, 4) the lack of pure autonomy without human control to be an effective deterrent, and the ennui of a large organizational bureaucracy that continues to promote the past instead of the future.

2.2 LMACC AND IMPLEMENTATION

How does a new concept that will meet the challenges of the Pacific and Indo-Pacific get through the myriad acquisition processes, and once there, find itself quickly able to be implemented into the Fleet? It is impossible to do justice here to the complexity of all the system of system inputs that produce a ship build. And further, is the ability to shift concepts addressing the problem of too few ships v. too many challenges. In many cases, a one-to-many np=complete problem that can't be solved in wartime. Here I will name some of what is necessary. First, as mentioned prior, a sense of urgency must be achieved. A sense of urgency, short of existential threat, must exist. In the throes of such a threat, it is to late. This sense of urgency must span across the U.S. Navy's funded PMWs, which will be (hugely) reluctant to shift to anything that could shake up their own budgets. Second, the system must be within the DON's budget. Third, the shift of fleetwide concepts must become a priority across the navy, fourth, the means to fund and produce a trial version of the concept must be undertaken (again with the sense of urgency). Fifth, the emphasis should be placed on "many" versus a "few." And sixth, the concept must be sustainable with the industrial base and its ability to add value under pressure.

In their very important study [5] Kline and Hughes write "Addition of a flotilla of small missile ships is recommended as a low cost, high reward component of the U. S. Navy. Emphasis is on the flotilla's complementary role in an affordable "Offshore Control Strategy" to maintain American influence with China and our Asian partners in the Western Pacific." LMACC has been designed around this concept, while also writing the concept of operations that allow it to become part of a sustainable flotilla (6 LMACCs and 6 unmanned vessels) the engineering specifications and other details have been provided in previous articles and are available at CIMSEC [6], RINA [7] and RINA [8]. These references fully explore the LMACC concept, its adaptability to the near and mid-term problems of the Indo Pacific, and provide design and capability specifics.

One area not fully covered in previous articles is that of the industrial base. To be effective at scale, the U.S. industrial base needs to be considered. That LMACC is not an esoteric hull and materials design to keep costs low, makes it possible for it to be manufactured in many current yards producing ocean going yachts. After the initial hull build, these hulls would be transferred to another facility where weapons, sensors and classified systems are installed. An article was published [9] that analyzed the inability of the U.S. Navy to depend on its own government shipyards for anything past refit and serious maintenance. LMACC creates the inclusion of civilian yards and greatly expands the industrial base.

3. REFERENCES

1. KLINE, J E., 'The "Bi-Modal" Naval Force Design Revisited', In *The U.S. Navy and the Rise of Great Power Competition. Routledge*, 2023.

2. GALLUP S P., 'Extended Thought, The Lightly Manned Autonomous Combat Capability (LMACC), *RINA* 2024.

3. KOTTER, J P., 'A Sense of Urgency.' *Journal of Technology Case and Application Research 2008-07-01, Vol 10* (3), pp 93-96

4. U.S. Naval Institute. *Pentagon Unmanned Systems Integrated Roadmap 2017-2042.* USNI News (2017).

5. KLINE, J E., and HUGHES, W P., 'A Flotilla to Support a Strategy of Offshore Control', *Technical Report, Naval Postgraduate School,* 2012.

6. DIDONATO, B., 'Lifting the Veil on the Lightly Manned Surface Combatant.' *Center for International Maritime Security(CIMSEC).* July17,2020. https://wp.me/p2moGg-bCa

7. GALLUP, S., 'Lightly Manned Autonomous Combat Capability Securing the World's Oceans.' *RINA 2023.*

8. GALLUP, S., et al., 'Expanded Thought: A Lightly Crewed and Highly Automated Warship Design.' *Future Warship Design 2024 Adelaide Australia, RINA proceedings.*

9. GALLUP S., et al., 'Impact of the Navy's 30-Year Shipbuilding Plan on the U.S. Industrial Base.' *Proceedings of the U.S. Navy Acquisition Research Symposium, Naval Postgraduate School* May 5, 2024.

4. AUTHORS BIOGRAPHY

Dr. Shelley Gallup holds the current position of Associate Research Professor at the U.S. Naval Postgraduate School within the Information Sciences department. With a distinguished career spanning both academia and military service, Dr. Gallup brings a wealth of experience and expertise that includes service as a Surface Warfare Officer (SWO) in the U.S. Navy. Dr. Gallup's research interests encompass the intricate domains of knowledge systems, organization development, and research methods applied to complex systems. Dr. Gallup's interdisciplinary background of biology, engineering (space systems), a graduate of the U.S. Navy War College and a career as a Surface Warfare Officer during the years spanning the end of the Vietnam war and the Cold War have contributed to a deep understanding of both military and academic contexts. She is uniquely positioned to address multifaceted challenges in current U. S. Navy policies. She has been the project manager of the LMACC project for the past 5 years and has written and been extensively published in journals, book articles (cybernetics and emergence in warfare) and conferences.

START WITH THE WHY – THE VALUE OF MARS COMPETENCE CENTRE

S Jensen, Svendborg International Maritime Academy, Denmark

SUMMARY

This paper examines the challenges of regulating emerging maritime technologies, particularly Maritime Autonomous Surface Ships (MASS). The rapid pace of technological advancement often outstrips the development of regulatory frameworks, creating a need for proactive and adaptive governance. The International Maritime Organization (IMO) has initiated efforts to address these challenges through a regulatory scoping exercise and develop a non-mandatory MASS Code. The paper highlights the importance of collaboration between private industry and public authorities to ensure regulations that enable innovation while safeguarding society. The Maritime Autonomous Reliable Systems (MARS) Centre is proposed as a potential solution, serving as a public competence hub to bridge knowledge gaps, support regulatory development, and foster societal readiness for new technologies. The MARS Centre can play a critical role in shaping effective and forward-looking maritime regulations by equipping regulators with the tools to ask the right questions.

NOMENCLATURE

HSE	Health Safety Exertive
IMO	International Maritime Organization
OOW	Officer On Watch
POLAR Code	International Code for Ships Operating in Polar Waters
MASS	Maritime Autonomous Surface Ships
MARS	Maritime Autonomous Reliable Systems
RSE	Regulatory Scoping Exercise
SIMAC	Svendborg International Maritime Academy
STCW	International Convention on Standards of Training, Certification and Watchkeeping for Seafarers
UNCLOS	The United Nations Convention on Law of the Sea

1. INTRODUCTION

'Every choice we make is beset with uncertainty.... A great deal of risk analysis is concerned with trying to turn uncertainties into probabilities' [1].

This quote illustrates the challenges legislators face in a world of rapid technological development. New and innovative technology always raises questions about safety, unknown risks, consequences of use, and how to protect the general population and workers. States usually try to solve these problems through regulations. The development of technology, though, is usually much faster than the development of new regulations. Regulators are faced with the legal dilemma of whether it is possible to contain new technology within existing regulations or if new legal instruments are required.

Within the maritime domain, much of the regulation that today governs both ships and maritime business derives from a time when ships were independent units, with little or no communication between ship and company while at sea and where navigation instruments did not provide the ability to navigate as precisely as today. One of the biggest challenges of the maritime industry is that the fleet is very heterogeneous, ranging from very modern and technically advanced vessels to vessels that have been operating in the market for 50 years and are equipped with old technology.

The so-called fourth industrial revolution with artificial intelligence and big data analytics is expected to contribute to transforming the shipping industry towards ever more autonomy. This prediction has caused a need to review the existing regulatory framework. Consequently, in 2017, the International Maritime Organisation (IMO) initiated a regulatory scoping exercise (RSE) to assess all existing IMO instruments and consider how they might apply to ships that utilise varying degrees of automation. Based on the RSE, developing a non-mandatory Code to cover regulatory gaps has been initiated.

One of the challenges the regulators face is regulating new technology that has yet to be developed. In the past, international maritime regulations have been reactive and often created in the aftermath of accidents. However, this time, regulators have chosen to be proactive, acknowledging that existing regulations otherwise risk becoming a barrier towards innovation.[2] However, this again raises the question of how regulators obtain the necessary knowledge to develop good and efficient regulations when new technology is often designed and tested by an industry that carefully guards its expertise due to the competitive advantage it constitutes.

In the best of all worlds, regulation enables innovation while simultaneously protecting society.[3] To obtain good and efficient regulation, a lack of access to data and knowledge can be a barrier, which can result in the limitation of innovation instead of enabling it. Therefore, there is a need for cooperation between the industry (the private actors) and the state (public actors) to create sufficient regulations. However, suppose the private entity retains the knowledge, and the state's role merely legitimises the industry standards. In that case, it may pose a risk to society due to possible different goals or perceptions and risk acceptance.[4] One way to bridge private actors and authorities could be a public competence centre, with capacity-building potential, enabling facilitation of regulators, supporting training and enhancing societal readiness. The centre could not provide all the answers; however, it will equip regulators with the necessary knowledge to ask the right questions.

In Svendborg, as part of the Svendborg International Maritime Academy (SIMAC), such a centre was established in 2022, denoted the MARS project. MARS stands for Maritime Autonomous Reliable Systems, and it has aimed to establish an international centre for testing, development, and production of advanced robots and reliable autonomous systems.

This paper focuses on the role of such centres to explain why a public competence centre, such as MARS, is beneficial. The paper departs from the presentation of risk-based regulations at the general level in Chapter 2. It follows an introduction to this approach towards the Regulation of Maritime Autonomous Surface Ships (MASS) in Chapter 3. Chapter 4 briefly describes the MARS project and Chapter 5 discusses the benefits of a centre such as MARS. The last chapter, Chapter 6, contains the conclusion and perspectives.

2. RISK-BASED REGULATION

'Not all risks can be reduced to zero, and trade-offs in risk reduction measures are inevitable' [5].

There is a constant strive for better regulation and a pursuit towards more efficient regulation, and one of the new governance techniques is risk-based regulation. Even though risk, as an object of regulation, is not new, neither is the fact that it legitimises the state's intervention in society. [6] Risk-based regulation and the state's role are ever contested. Some are concerned with a shift towards what they call 'the nanny state' that ties individuals and companies into 'red tape' and takes away their freedom and personal responsibility for their own choices. Others argue that the state needs to protect society more and that there is a need for further regulation.

One of the challenges of risk-based regulation is the definition of risk and who defines what is perceived as a risk. According to Giddens, risk has evolved from *external risk* to *manufactured risk.* [7] E*xternal risks* are events that strike individuals from the outside with a regularity that makes them predictable and insurable. *On the other hand, manufactured risks* are created primarily by the progression of science and technology without knowing their consequences. The probability and consequences of those risks are ever-changing, developing and increasing. Giddens uses smoking as an example: in the 70ies, doctors recommended smoking as a means of relaxation, but since the end of the 90ies, the perspective on smoking has changed, and it has become a risk that is increasingly regulated. Smoking is probably still relaxing, and health issues related to smoking were also known in the 70ties, but the focus has changed. Studies of long-term effects on health are now publicly known, and many are familiar with the long-term consequences, knowing people who have died of illnesses related to smoking. The long-term health effects of smoking have put pressure on healthcare systems and have had significant economic consequences. However, smoking is not only a burden to the state but also a source of income through cigarette taxes and the tobacco industry. This increases the complexity of the state's choice of measures, as it must balance the interests of the public and the industry. The state faces a push-pull dynamic: some warn against scaremongering and argue for individual freedom of choice. In contrast, others accuse regulators of cover-ups to serve the industry and call for consistent action.

What constitutes the perception of risk and how to handle it is culturally conditioned, dependent on one's position and tied to ideology. How risk is perceived differs among groups and individuals; it depends mainly on their access to information, familiarity with the matter of issue, and trust in the ones who manage the risk.

When risks are communicated, it is often in terms of probabilities and percentages. Still, to many, this type of information is ambiguous, and how it is perceived again depends on familiarity with the issue and trust in the ones who manage the risk. As an example, one in 10.000 risk experiencing respiratory problems that could be life-threatening after ingestion of paracetamol, a risk that many people run every day when they, due to light pains, ingest a pill containing paracetamol. The risk 1/10000 is vague and ambiguous to many, and if the product is sold in grocery shops (in Denmark, it is), it signals to the individual that it is considered safe to use by the state, which is why most don't care to read the information list. Through its regulation, the state risks legitimising unsafe tasks, which has been of great concern in different cases. In the maritime domain, adopting the POLAR Code has brought forward the question of legitimising unsafe tasks, specifically concerning passenger vessels operating in Greenlandic waters [8]. Is it safe to sail a passenger ship along the coast of Greenland? The ship is complying with international rules and regulations; the flag state Administration verifies it, and as such, the passengers have a sense of safety. However, operating in harsh areas with limited rescue facilities always poses a risk. The POLAR Code sets up requirements the company shall comply with, but is it safe to operate in these areas? It is a matter of perspective. This illustrates a paradox of regulation: despite the aim to reduce risk, regulation can inadvertently lead to greater risk by increasing risk-taking activities.

2.1 TOLERABLE RISK

The risk-based approach forces the regulators to ask: How safe is safe enough? Safe enough is also denoted as the *tolerable risk.* A formal risk assessment requires calculating outcomes and probabilities, both of which will be uncertain for new technologies. Risk does play a significant role in regulation; however, the economy is still dominant. Thus, the question of tolerable risk becomes a trade-off between pros and cons. Consequently, the definition of a tolerable risk is rather political than scientific. It depends mainly on the regulator's perspective on what is a tolerable risk to them. It is important to note that tolerable risk is what determines the regulation, not the acceptable risk. Hence, an acceptable risk indicates that the person subject to a risk has accepted the risk. However, to accept a risk requires individual competence and knowledge. A tolerable risk indicates that the risk assessment concluded that the trade-off relative to the risk by the regulators was found to be tolerable. As the United Kingdom's Health Safety Exertive (HSE) established, tolerable does not mean acceptable. Instead, it refers to a willingness by society, as a whole, to live with a particular risk to secure certain benefits in the confidence that the risk is worth taking and that it is being adequately controlled. [9]

2.2 PRECAUTIONARY PRINCIPLE

Giddens stipulated that 'We increasingly live on a high technological frontier which absolutely no one completely understands' [7], a statement that challenges the definition of tolerable risk. Thus, it is necessary to touch upon the *precautionary principle* in the debate on risk-based regulation. According to the EU Commission, the precautionary principle should be essential to decision-makers. Consequently, when determining what constitutes an acceptable risk, they must recognise scientific uncertainties and public concerns; their duty is to find answers. [10] This statement from the EU Commission aligns with concerns raised by the UK Commission on Better Regulation, which suggests that governments poorly understand trade-offs and costs and that this lack of knowledge affects the quality of regulation [11]. The precautionary principle is closely linked to environmental protection and calls for the prevention of severe and irreversible damage to the environment. Relative to risk-based regulation, the precautionary principle forces the regulators to act. Despite being a core principle to environmental and risk regulation in the EU, the principle is also contested, with opponents arguing that the *precautionary principle* acts as a barrier to innovation, delaying the introduction of new technology and raising costs.[12]

3. REGULATING MARITIME AUTONOMOUS SURFACE SHIPS

The fourth industrial revolution, where artificial intelligence and big data analytics are expected to transform the shipping industry and introduce smart autonomous ships, is highly prioritised on the IMO agenda. As stipulated in the introduction, IMO initiated an RSE to determine regulatory gaps and barriers, and based on those results, the member states decided to develop a non-mandatory Code aimed at regulating smart autonomous ships that are unable to comply with conventional regulations. IMO, denote these vessels as Maritime Autonomous Surface Ships, abbreviated MASS. They have defined four types of MASS based on their degrees of autonomy. Those are:

1. Ships with automated processes and decision support. Operations may be automated and, at times, unsupervised, but with seafarers onboard ready to take control.

2. A remotely operated ship with seafarers onboard. The ship is controlled and operated from another location, but the seafarers are onboard, ready to take control and able to operate the ship.
3. Remotely controlled and operated vessel, with no seafarers onboard.
4. Fully autonomous ship. The operating systems can make decisions and determine actions without human interference.

It is important to note that the IMO notion is not related to levels of autonomous ship design and operation, as Lloyd's Register has defined it [13], but to different classes of vessels.

The RSE was conducted based on those four types, exploring whether a type is expected to be able to comply with a specific regulation. MASS 1 is more or less the type of vessel that is in operation today; onboard those vessels, some operations are automated and unsupervised but with a human in the loop, e.g. an autopilot that keeps the ship on course, giving commands to the steering machine, with an alarm that will warn the officer on watch (OOW) in case of irregularities. MASS 3 and 4 are the ones that challenge the regulation the most. It is recognised that increased digitalisation, interconnection and advanced technologies have the potential to enhance safety, efficiency and environmental protection. However, there are still many uncertainties. IMO's Strategic plan for 2024 to 2029 [14] to *integrate new, emerging and advancing technologies in the regulatory framework* recognises that technological advances represent opportunities. Meanwhile, it emphasises the need to balance the benefits of maritime safety and security concerns, cybersecurity, environmental concerns, potential for international trade facilitation, costs, and the impact on personnel, both on board and ashore.

3.1 IDENTIFYING THE RISK

As mentioned before, introducing smart autonomous ships represents benefits and various risks. A key concern is that MASS challenges the global governance system constituted by IMO. According to IMO's strategic plan, 'the main role of IMO as the global regulator of safe, secure, environmentally sound and efficient shipping requires it to ensure that a universally adopted, effective, international regulatory framework is in place and implemented consistently, embracing and integrating new, emerging and advancing technologies, without causing unnecessary burden.' [14, para 46]. Later, it is specified that by developing regulation and goal-based standards, IMO seeks to ensure a level playing field.

IMO regulations and the United Nations Convention on Law of the Sea (UNCLOS) are considered barriers to accepting MASS 3 and 4 ships operating on high seas. For example, according to UNCLOS article 94.4(b), a ship shall be in charge of a master and officers. However, UNCLOS does not define the location of the master and officers; this is an IMO requirement as the master and officer on watch, according to the STCW convention, are required to be onboard the vessel. When the ship, on the other hand, operates in territorial waters, the state may grant an exception. Yet, this exception only applies if the ship operates in waters subject to state sovereignty. If the ship needs to enter another state's territorial water, it must be granted an exception from that particular state. In contrast, if a vessel complies with IMO regulation, it is granted free passage through other states' territorial waters if its intentions are peaceful. This secures the playing level field, eases the trade and prevents discrimination. Under the Convention on IMO, removing discriminatory actions and unnecessary restrictions is a fundamental principle. So is also a principle based on a more traditional perception of risk. Namely, to ensure the highest practicable standards in matters concerning maritime safety, environmental protection, and navigation efficiency.[15] Those are the two principles the non-mandatory MASS Code must adhere to. Until the MASS Code is in place, approval guidelines have been adopted to cover advanced technology. Those are explored in the following section.

3.2 GUIDELINES FOR THE APPROVAL OF ALTERNATIVES AND EQUIVALENTS AS PROVIDED FOR IN VARIOUS IMO INSTRUMENTS

The development of the MASS Code is in progress, but in the meantime, the Maritime Safety Committee (MSC) Circ. 1455 'Guidelines for the Approval of Alternatives and Equivalents as provided in various IMO Instruments' applies to MASS. The guideline is based on the concept that innovative design can be accepted if the safety level is equivalent to that of a conventional design. Circ. 1455 defines *safety* as the absence of unacceptable levels of risk to life, limb and health.

Consequently, as deVos and Hekkenberg posit, 'something is safe as long as it is subject to an acceptable risk' [16]. Risk is also defined as a concept in circ. 1455 para 2.14 here it states that '*Risk* is a measure of the likelihood that an undesirable event will occur together with a measure of the resulting consequences within a specific time, i.e. a combination of the frequency and the severity of the consequences (this can be either a quantitative or qualitative measure).' This leads us back to the statement in Chapter 2 on Risk-based regulation, which started with, 'risk cannot be reduced to zero, it is a trade-off'; hence, the acceptable level of risk is a political decision, balancing pros and cons, determining the tolerable risk. Thus, the guideline describes the process, yet it is up to the Administration to decide whether or not the risk is tolerable.

One approach described by circ. 1455 is the demonstration of equivalence. It requires that functional requirements and performance criteria are established for the essential ship functions, which should be met by the alternative design. The guidelines pose a structured approval process. In Denmark, the Regulatory Future Lab facilitate the approval process based on Circ. 1455. An essential part of this process is to define the acceptance criteria, as those will be the basis for the approval; the applicant is required to provide a report that includes a description of the design, reference to the regulation affected, safety hazards, risk assessment and proposals for mitigating measures. Depending on the preliminary report's risk and quality, the Administration could request further mitigating measures and tests before final approval.

Circ. 1455 emphasises that 'the Administration needs to be in a position to assess whether the design has been sufficiently examined and whether any risks have been reduced acceptably and thus should have sufficient knowledge of the subject to evaluate the adequacy of the delivered information and the assumptions made'.

The acceptance criteria are determined by the Administration, in Denmark, the Danish Maritime Administration (DMA). Coglianese and Mendelson argue that regulators are likely to find themselves at a significant information disadvantage in cases of complexity and emerging risk compared to the industry they oversee. [17] The Administration poses an in-depth knowledge of the regulation that conventional ships are to comply with, a regulation that is predominantly reactive, constructed to avoid known consequences, and primarily based on command-and-control principles. But what is sufficient knowledge concerning innovative and alternative design? How does the DMA obtain expert knowledge? The developers possess that knowledge; still, the DMA must be able to set acceptance criteria, evaluate the information, and ask the right questions.

3.3 EXPERIENCES FROM SELF-DRIVING CARS

The development of autonomous shipping is often compared to autonomous cars, arguing that self-driving cars are soon to become a fact of life. [18] However, self-driving cars are struggling to demonstrate a tolerable level of safety.
Most safety case approaches are performance-based, based on a tradition for safety-in-number, meaning that to document safety, it is a question of the number of kilometres the car drives without incidents. This number is then compared with a conventional car. The safety-in-numbers approach is subject to several uncertainties. This extensive focus on performance displaces questions on process and purpose concerning self-driving cars. Thus, those are largely neglected by the industry. Therefore, there is a call for the industry to open up the process and allow for looking into the technological black boxes. The regulators need not only to accept developers' questions as sufficient. Studies call for attention to turn the process upside down, accept that things go wrong, and then find out everything is wrong with the system instead of focusing on performance alone.[19]

The call from the car industry to focus on the wrongs is reminiscent of what in shipping has been denoted as near-miss reporting. For the past decade, this approach has been of key value to shipping; it forms a breeding ground for learning and knowledge on how processes can be improved. It is in contrast to the safety-in-number approach, which results in a probability that, like paracetamol, as mentioned above, can be difficult to relate to. Quantitative studies must be supplemented by qualitative processes, which can be performed at a test centre.

4. THE MARS PROJECT

The MARS project is part of NextGen robotics and is funded by the EU. The overarching vision is to establish an international centre for testing, developing, and producing advanced robots and reliable autonomous systems designed for land, air, and marine applications. Specifically, the maritime aspect of the project is situated in Svendborg as part of Svendborg International Maritime Academy (SIMAC).

At SIMAC, future Marine Officers receive training, preparing them to operate alongside and collaborate with cutting-edge technology. The coastal area near Svendborg serves as a testbed for advanced technology, and the MARS Centre maintains an equipment bank. These resources also allow minor players and students with limited resources to initiate and participate in innovative processes and conduct tests.

The MARS project's vision is clear: to be the hub where autonomous solutions can be tested, regardless of whether the stakeholders are entrepreneurs, established companies, research institutions or educational organisations. Danish and foreign companies can benefit from being part of a knowledge environment that fosters joint innovation, collaborations and networking.[20] The MARS project aims to support private-public partnerships, and as such, the advisory board consists of representatives from industry, training institutions, and universities, and a close dialogue with public authorities
supplements this.

The MARS centre may be denoted a Centre of Excellence based on the above description.[21] A Centre of Excellence is an environment that fosters research, innovation, and learning, allowing the ability to absorb and generate new knowledge, which fits the overall aim of the MARS project.

5. THE WHY

This article explores why a public competence centre like MARS is beneficial. The introduction of autonomous systems in shipping is presumed to bring benefits, such as efficiency, reduced human errors and enhanced safety [22]. Still, as argued above, these are the presumed benefits, but the pros seldom come without the cons. Industry has set out to explore and develop smart autonomous systems for shipping; those are developed and tested in closed areas, and information, data, and learning accessible to the public is limited. In Norway, Yara Birkeland, a container vessel expected to operate as a MASS3 in the future, is currently being tested as an MASS2 vessel. Sea Machines are testing in Boston Bay, and the Nippon Foundation is behind the test of a MASS container vessel in Tokyo Bay. [23] Those are examples of MASS tests, whether they are still national or private tests, and they are few. In Denmark, the Green Hopper [24] is one of the few concepts that has been tested. However, the project has been closed down and only managed to operate with a crew onboard. Learning from this project is still valuable, and that is why people involved in that project have also been invited to take part in MARS. Also, a private entity such as Sea Machine is part of the advisory board to ensure synergy with other projects.

Risk-based regulation, tolerable risk and acceptable risk are closely linked to probability or the safety-in-number tradition applied by the self-driving-car industry. However, acceptance based on calculating probability is challenging. As Eriksen and Lützen [25] stipulate, a probability of failure of one in 190 years for a pump operating in one subsystem needed for propulsion might be found to be tolerable, but there are several essential subsystems. Thus, the cumulative probability must be expected to be much higher. Consequently, the actual risk of failure is difficult to foresee. Redundancy is one way to reduce probability. Still, Eriksen and Lützen argue that 'redundancy may not be enough because dependent failures effectively limit the effectiveness of redundancy on reliability.' Data on incidents and accidents on conventional ships are collected. However, the incidents resulting in an accident are only considered the tip of the iceberg. It is recognised that in many situations, the onboard crew can prevent an incident from becoming an accident. Consequently, the likelihood of an undesirable event occurring thus can be difficult to compare for manned and unmanned vessels.

Relative to self-driving cars, there was a call for opening up the black box of technology to lay out all the wrongs instead of counting the rights. By testing, collecting data, and facilitating research, the MARS centre has the ability to open the black box. I have divided it into three core values: Capacity building, training and education, and societal readiness, each explained below. The figure below (figure 1) illustrates the network of competencies for which MARS has the potential to be the centre by becoming a Centre of Excellence and cooperating with private actors (the industry) as well as public actors (the DMA), influencing the maritime education and training and increasing knowledge in society as whole.

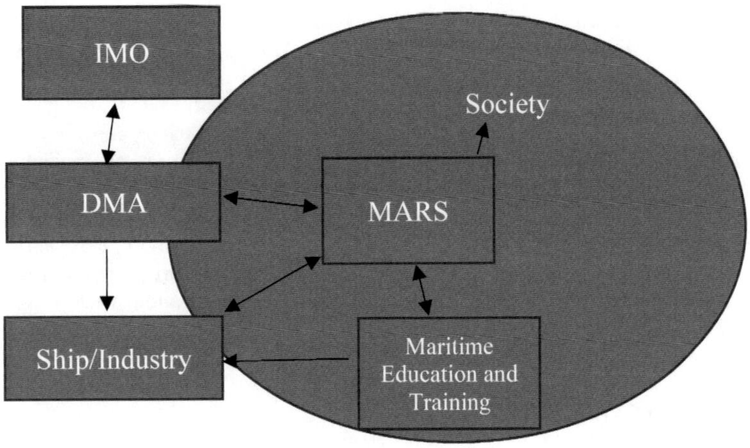

Figure 1: MARS Centre of Excellence network

5.1 CAPACITY BUILDING

One of the key potentials of the centre will be capacity building through cooperation between actors and gathering knowledge. This capacity enables qualified political decisions relative to MASS, acceptance criteria, evaluation of information and ability to ask the right questions. Testing of innovative designs, gathering of data, and cooperation with researchers, posts and undergraduates has the potential to open the black box of technology; it will not offer a complete understanding of new technology but enhance the knowledge of new technology, the pitfalls and need for mitigating measures, but also the benefits. Thereby avoiding unnecessary barriers but also that free market forces determine the course due to a lack of knowledge.

5.2 TRAINING AND EDUCATION

As stated in Chapter 3, smart autonomous systems will transform shipping in the coming decade. Still, a human in the loop is expected to be necessary, whether onboard or onshore and those will be trained at maritime academies. As part of SIMAC, where future seafarers are trained, the MARS centre will ensure close cooperation between seafarers and the systems' developers. This synergy between SIMAC and MARS is valuable. Knowledge gained at the MARS centre can be implemented into the training programmes so that the students may be part of developing and testing the new technology. The cooperation has the potential to ease the implementation of new technologies, but it will also qualify the process, with seafarers becoming part of the innovation and testing. The implementation phase is a primary risk of new technology; being part of a MET centre will mitigate the risk. The training is often based on past technology; working together with MARS will increase the knowledge of tomorrow's work processes and could be included in the curriculum.

5.3 SOCIETAL READINESS

'Societal readiness refers to society's receptiveness to a new technology and readiness to accept this technology's use in the public domain.' [26] As posited above, regulation is based on a tolerable risk, a political definition of what risk society can accept. However, this does not mean an individual finds the risk to be acceptable. As noted, knowledge and trust influence the perception of a risk. According to Knudsen and Kiel [26], the MARS project can potentially raise Societal Readiness by conducting initial and pilot tests. Enhancing the knowledge level and being a public Centre of Excellence working with private entities, the MARS project may also counteract the sense of industry cover-up, thereby supporting trust in society. Knowledge and trust are vital to raising societal readiness.

6. CONCLUSIONS

As we enter the fourth industrial revolution, with artificial intelligence and big data analytics transforming the shipping industry, smart autonomous ships are being tested worldwide. Nevertheless, it is essential to accept that despite the word revolution, which means a sudden and significant change, there are many small steps towards the all-encompassing transformation. Technology and regulation are evolving, and competencies and education must evolve with it. A centre of excellence such as MARS, with its unique location as part of a maritime education and training centre, allows for capacity building in close collaboration with the people who are to work alongside the systems; this will qualify the system development as well as the education. Enhancing the man-machine interface, creating a synergy, and increasing the students' knowledge about and trust in the new technology.

Risk cannot be depoliticised; what constitutes a risk and what is considered a tolerable risk is de facto a political decision. However, by opening up the process to the public and the regulators, risk governance becomes democratic. Transparency and an open test environment increase trust in the technology and the regulators, which is necessary for public acceptance (Societal readiness). Some might say that public acceptance is not required in the same way that it is for self-driving cars, but studies have shown that if an accident is related to one type of transport, it affects the public trust across means of transport. [26]
Finally, the MARS centre can facilitate public-private cooperation where developers can test their systems. Data, information, and learning are collected, stored, and analysed by a neutral actor, and learning from this centre can support the development of efficient regulation, qualifying the regulators' questions.

7. ACKNOWLEDGEMENTS

This article was written as part of the Danish MARS project (Maritime Autonomous Reliable Systems), which is funded by the EU Social Fund and the EU Regional Fund. Thanks to MARS project leader Charlotte Kirkegaard Flugt and senior researcher at CMSS Hanna Barbara Rasmussen for professional sparring.

8. REFERENCES

1. DOUGLAS, M., 'Risk Acceptability According to the Social Sciences', *London: Routledge & Kegan Paul*, 1985. (page 42)

2. IMO, 'MSC. 1/Circ. 1638. Outcome of the Regulatory Scoping Exercise for the use of Maritime Autonomous Surface Ships (MASS)', London, 2021.

3. WEIMER, M., and MARIN, L., 'The Role of Law in Managing the Tension between Risk and Innovation', *European Journal of Risk Regulation Vol. 7, No. 3*, 2016.

4. COGLIANESE, C., and MENDELSON, E., 'Meta-Regulation and Self-Regulation', in Baldwin et al. (eds), *The Oxford Handbook of Regulation,* 2010.

5. OECD. 'Risk and Regulatory Policy - Improving the governance of risk', *OECD Reviews of Regulatory Reform*, 2010.

6. BLACK, J., 'The Role of Risk in Regulatory Processes', in Balwin et al. (eds) *The Oxford Handbook of Regulation*, 2010.

7. GIDDENS, A., 'Risk and Responsibility', *The Modern Law Review, Vol 62, no. 1*, 1999.

8. RASMUSSEN, H B., and JENSEN, S., 'The Polar Code vs The International Safety Management Code: Do We Need Both?' in *The United Nations Convention on the Law of the Sea: A System of Regulation* (s. 223-237). Routledge, 2024.

9 HSE, 'Reducing Risks, Protecting People - HSE's Decision-making Process', Health and Safety Executive, 2001.

10. European Commission, 'Communication from the Commission on the precautionary principle', COM/2000/0001 final, 2000.

11. Better Regulation Commission (BRC), 'Risk, Responsibility, Regulation: Whose Risk is it Anyway', *London: Cabinet Office*, 2006.

12. AVEN, T., 'A risk and safety science perspective on the precautionary principle', *Safety Science (165),* 2023.

13. BLANKE, M., HENRIQUES, M., and BANG, J., 'En for analyse om autonome skibe', *DTU,* 2017.

14. IMO, 'Resolution A.1173(33) Strategic Plan for the Organization for the Six-Year Period 2024 to 2029', London, 2023.

15. IMO, 'CONVENTION ON THE INTERNATIONAL MARITIME ORGANIZATION' as amended

16. DE VOS, J., and HEKKENBERG, R., 'Towards Safety Regulations for the Design of Autonomous Ships', *COMPIT 2020*, 2020.

17. COGLIANESE, C., and MENDELSON, E., 'Meta-Regulation and Self-Regulation', in *Robert Baldwin, Martin Cave, and Martin Lodge (eds), The Oxford Handbook of Regulation,* 2010.

18. SEA MACHINES, 'Why Reinvent the Maritime Wheel?', 2021.

19. STILGOE, J., 'How can we know a self-driving car is safe?', *Ethics and Information Technology 23:635-647*, 2021.

20. According to the MARS webpage: https://marsdenmark.dk/om-projektet/om-projektet, 2024.

21. HELLSTRÖM, T., 'Centres of Excellence and Capacity Building: from Strategy to Impact', *Science and Public Policy, 45(4),* 2018.

22. SIVORI, H., and BRUNTON, L., 'Out of the Box – Implementing autonomy and assuring artificial intelligence in the maritime industry', LR, 2024.

23. EWING, T., 'Autonomous Vessels: Already Working, but Still Under Close Review', *MarineLink,* Juni 17, 2024.

24. DITTMANN, K., and BLANKE, M., 'Risk mitigation by design of autonomous maritime automation systems', *at – Automatisierungstechnik 70(5),* 2022.

25. ERIKSEN, S., and LÜTZEN, M., 'The impact of redundancy on reliability in machinery systems on unmanned ships', *WMU J Marit Affairs 21, 161–177*, 2022.

26. KNUDSEN, M P., and KIEL, J., 'Society's Readiness for Autonomous Shipping' in *JENSEN et al. (eds) A framework for understanding the barriers when sailing autonomously*, 2023.

9. AUTHORS BIOGRAPHY

Signe Jensen is a Postdoctoral Researcher at the University of Southern Denmark (SDU) in the Faculty of Law. With a robust academic and professional background, Signe has also held the position of Associate Professor at the Svendborg International Maritime Academy (SIMAC) until May 2024.

Signe's expertise bridges the fields of maritime operations and environmental law. She holds a Master Mariner degree, demonstrating her deep understanding of maritime practices, as well as a Master in Environmental and Energy Law. Her academic journey culminated in a PhD focused on a sociolegal study of the International Safety Management (ISM) Code, where she explored the intricate relationships between legal frameworks and maritime safety practices.

Signe's multidisciplinary research combines her practical experience in the maritime industry with legal analysis. This unique perspective informs her contributions to the field, particularly where law and maritime safety intersect.

DANELEC AND DANPILOT REMOTE PILOTING

T Petersen VP, Safety Product Management at Danelec, Denmark

SUMMARY

Danelec and piloting companies are advancing remote piloting technology for maritime navigation. This system lets pilots monitor ships from a control center, reducing risks, boosting efficiency, and lowering costs. It cuts fuel consumption and emissions by removing the need for physical pilot transfers, ensuring safer and more eco-friendly maritime operations.

1. INTRODUCTION

DanPilot and Danelec are collaborating on remote piloting initiatives. This technology ensures safer, greener, and more efficient maritime operations. For ship owners this means:

- Minimized crew interaction
- Faster voyage
- Less Fuel Consumption (Greener)

For pilots the advantages are:

- Avoid hazardous ship transfers
- Weather independent
- Simpler planning

2. NAVIGATION SYSTEM OVERVIEW

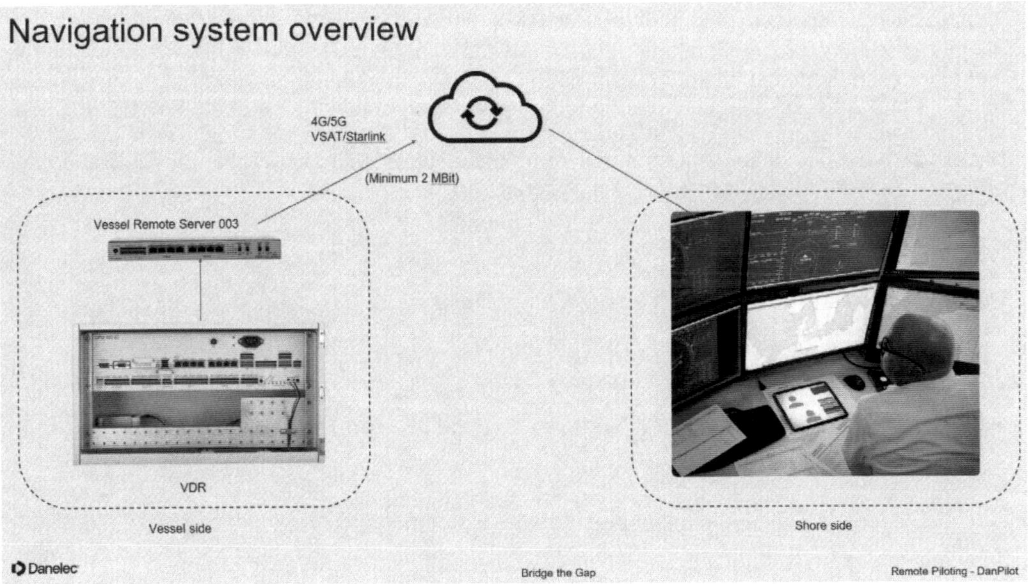

Figure 1: Navigation data

3. COMMUNICATION SYSTEM OVERVIEW

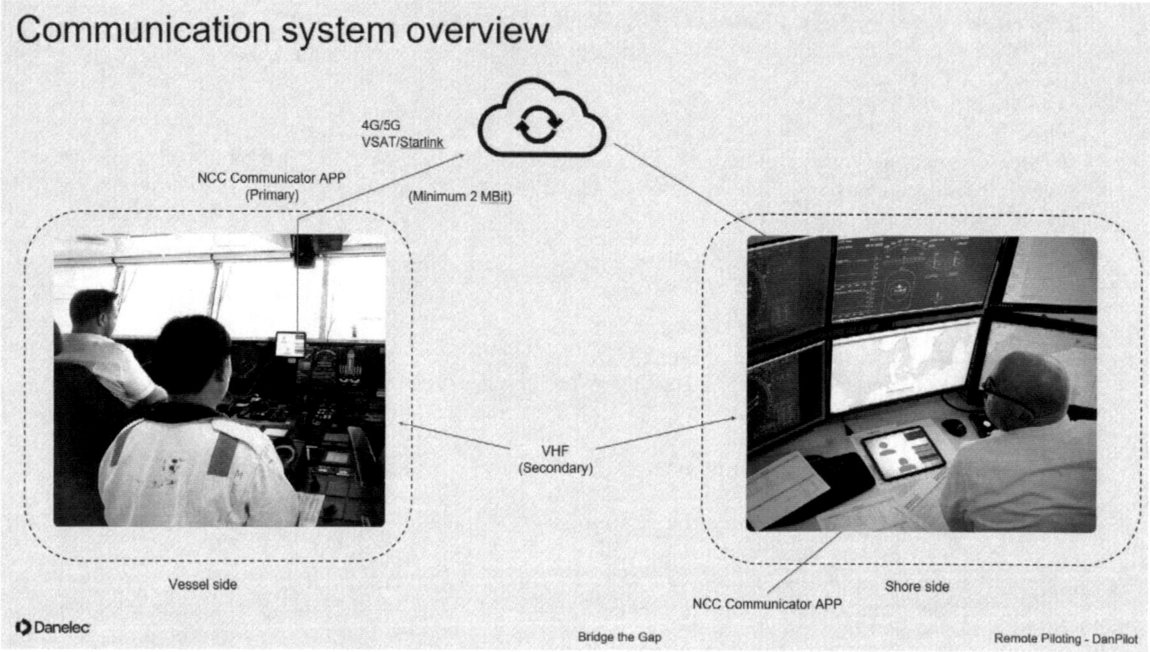

4. REQUIREMENTS

For a remote piloting session two conditions need to be present. A) the right vessels B) the shore setup. Once that is in place:

5. CONCLUSIONS

There is still some time before the test are finalized and the regulation is in place, however remote operations will become a viable alternative for traditional piloting soon.

6. AUTHORS BIOGRAPHY

Thomas Petersen VP, Safety Product Management at Danelec.

THE ROYAL INSTITUTION OF NAVAL ARCHITECTS

THE ROYAL INSTITUTION OF NAVAL ARCHITECTS

8-9 Northumberland Street, London WC2N 5DA

Tel: +44 (0)207 235 4622

events@rina.org.uk

www.rina.org.uk

Stay connected with our social platforms